SUSTAINABILITY REPORTING BY MINING COMPANIES IN GHANA

SUSTAINABILITY REPORTING BY MINING COMPANIES IN GHANA

Clement Lamboi Arthur

To order additional copies of this book, contact:
Xlibris
800-056-3182
www.Xlibrispublishing.co.uk
Orders@Xlibrispublishing.co.uk
777536

TABLE OF CONTENTS

LIST OF TABLES

PREFACE

Environmental, economic and social sustainability management is one of the most complex challenges facing both private and public sector organisations in recent times especially with respect to reporting. The book is intended to demonstrate the kind of indicators mining companies in Ghana disclose in the sustainability reports. This book is different from other accounting books in that unlike other texts on accounting book, it addresses the needs of variety of users in spite of the fact that relates Ghanaian mining companies. The book is intended to demonstrate the kind of indicators mining companies in Ghana disclose in the sustainability reports. Another distinction is that it meant for an all accounting audience who are interested with current issues in sustainability accounting and reporting of mining companies in Ghana. It provide a good starting point for anyone wanting to embark on research in this area. The book is meant to be easy to read so I apologise for including references and citations. If I left them all out, you would not know whether what I was saying was my personal opinion or had some authoritative evidence to back it up. I have tried to minimize the references though, consistent with the requirements of the need to evidence statements and give credit to innovative thinkers. I hope I have made it very obvious when anything is just my personal opinion. The content of this book is intended for practicing accounting and finance professionals, graduate students, and advanced undergraduate students. The social and economic context of sustainability accounting is regularly brought into discussion and accounting, like all of business studies is a social study. It is done by people about people to people. It is not just about what things people own and what those things are worth.

Acknowledgements

I would like to thank Almighty God for His protection and guidance throughout the writing of this book. The completion of this book has been the most challenging but desirable activity. I will want to thank the reviewers of this book Dr Junjie Wu in the Faulty of Accounting and Finance of Leeds Beckett University, Leeds, UK, who provided a comprehensive review of the book and edited by Dr Milton Yago in the Faulty of Accounting and Finance of Leeds Beckett University, Leeds, UK.

CHAPTER 1

Overview of Mining in Ghana

1.1 Introduction

Chapter provides an overview and historical review of the mining operations in Ghana. It is used to present the overview of global gold industry up to 2016. The chapter also used in presenting the history of mining in Ghana.

1.2 An overview of Ghana

Ghana is a West African country with a land size of 238,540 km2 (92,100 mile2), and it is situated on the Gulf of Guinea, an arm of the Atlantic Ocean (IFAD, 2007), which is comparable in size to the state of Oregon or about the size of United Kingdom (Assenso-Okofo et al., 2011). Ghana is located on the West African coast, bordered by Côte d'Ivoire to the west, Burkina Faso to the north, Togo to the east and the Gulf of Guinea to the south. It was formerly known as the Gold Coast, because of the abundance of gold in West Africa. Ghana has an estimated population of 25 million, which has been growing at about 2.7% per year (Gallardo, 2002; NDPC, 2005; GSS 2012). GDP per capita (PPP) in 2006 was estimated at $2700 (CIA, 2008). It is believed that more than 40% of the country occupies the Volta River Basin, a large area flooded by Lake

Volta where a dam was built (Akosombo Dam) to provide hydroelectric power for smelting alumina into aluminium (Berry, 1995; Boateng, 1967; Assenso-Okofo et al., 2011). The country has been divided into six main agroecological zones (coastal savannah, rain forest, deciduous forest, transition, Guinea savannah and Sudan savannah). The annual rainfall ranges from a low of 800 millimetres in the coastal savannah to a high of 2200 millimetres in the rain forest.

Ghana became the first country in sub-Saharan Africa to gain independence on 6 March 1957.The population living in rural areas has reduced from 63% in 1999/2000 to 53% in 2005 and there are indications that the rate of rural-urban migration will increase (Gallardo, 2002). The British took control of the country in 1821, and the colonization continued until Ghana attained its independence in 1957. After leading the country for 9 years, the nation's founding president, Kwame Nkrumah, was overthrown in a coup d'etat in 1966 (Assenso-Okofo et al. 2011). Subsequently, Ghana was ruled by a series of military despots with intermittent experiments with democratic rule, most of which were curtailed by military takeovers. The latest and most enduring democratic experiment began in 1992 and has led to Ghana's reputation as a leading democracy in Africa (Ghanaweb, 2009). Ghana adopted a parliamentary democracy at independence, and now in the Fourth Republic with presidential and parliamentary elections in 1992. In recent parliamentary and presidential elections held in December 2012, Mr John Mahama was elected president of Ghana, and his party, the national Democratic Congress (NDC) won a majority in parliament. The main opposition party is the National Patriotic Party (NPP).

The perception of corruption has been cited as the country four violent overthrows of governments by the military. In 1992 Ghana returned to multi-party democracy and has not departed from that form of governance. At independence, Ghana emerged as one of the biggest owners of reserves in the Sterling Area system due to the boom in the market for cocoa. Accumulated national savings led to a spate of infrastructure construction, including the Tema harbour, the industrial city of Tema and the Volta River power system.

Overall, economic growth was relatively robust after independence till the mid-1960s. The Ghanaian economy became sluggish from 1964 when the exhaustion of financial reserves coincided with the collapse of market for cocoa in the 1960s. Danquah (2006) observed that during much of the period between 1966 and 1983 the economic growth rates were negative.

Economic growth was uneven and poor after the mid-1960s and only began to stabilize after 1984. The economic improvement was due to the fact that in 1983 Ghana adopted the World Bank's prescription for development in Third World countries: the Structural Adjustment Programme (SAP), known in Ghana as the Economic Recovery Programme (ERP). The SAP involved policies that were meant to improve resource allocation, increase economic efficiency and improve the country's ability to withstand domestic and global economic shocks. From 2000 Ghana experienced a spectacular reduction in inflation rates and interest rates have somewhat dropped and regarding GDP, since 2000 real GDP rates have witnessed consistent growth, culminating in a 7.5% growth in 2008.

Ghana is located in a region with rich biodiversity, of national as well as global significance, including species that are endemic (unique) to the region. The country is part of the Upper Guinea forest ecosystem, which is categorised as one of the biodiversity hotpots in the world. As important as biodiversity values are, Ghana's ecosystems support human livelihoods and economic sectors. The natural resources (e.g. soils, water, minerals, forests, wildlife) are essential for productive sectors of Ghana's economy. In addition, the environment performs at no cost several regulating functions (e.g. soil protection, pest control, forest production, water purification, etc.).

However, the functioning of these productive and regulating functions depends upon proper environmental governance and management. Failing to do so, other (capital) resources will have to be used in order to compensate for the loss of environmental functions (e.g. bottled water if natural streams become polluted). In addition, a poorly managed environment will cause pollution of soils, waters and air which affects human health. Around 11 million of Ghana's population live in forest areas, and around two thirds of them are supported by forest-related activities (CEA, 2006). Yet, Ghana suffers from rapid deforestation and destruction of biodiversity. Apart from being the world's second largest cocoa producer the country extensively exports significant quantities of gold, timber, diamond, bauxite, and manganese. Ghana remains somewhat dependent on international financial and technical assistance, as well as the activities of the Ghanaian Diaspora. Recent oil finds in 2007 reported to contain up to 3 billion barrels (480,000,000 m3) of light oil has engendered optimism that the country can reach middle-income status by 2020. Oil exploration is on-going and the estimates of oil reserves in the fields continue to increase. The domestic economy revolves around subsistence agriculture, which accounts for 50% of GDP and employs 85% of the workforce.

1.3 Overview of the Global Gold Industry in 2016

According to Ghana Chamber of Mines (2016) in the beginning of 2016, the bullion market responded to the United States unexpected weak 2015 fourth quarter economic data with a surge in price. These data compelled most investors to revise their expectations about a possible hike in the Federal Reserve's fund rate. Consequently the investors reorganize their portfolio investments in favour of the gold. This action affected, the price of gold which rose from USD 1,082 per ounce to USD 1,277 per ounce in March on the London Mercantile Exchange. The downward and upward movements in the price of gold in the month of April, caused the traded price to decline to USD 1,212 per ounce on 2nd June, 2016. However, the Brexit uncertainty triggered a bullish run of the gold price, to bring to highest price of USD 1,366 per ounce in July. Overall, the cumulative average price of the yellow metal in 2016 was 1,250 per ounce. This represents a 7.75 per cent increase over the cumulative average price of USD 1,160 per ounce recorded in 2015.

Table 1.1 Top 20 Gold producing countries production (tons)

Countries	2011	RANK	2012	RANK	*2013	RANK	2014	RANK	2015	RANK	2016	RANK
South Africa	197.9	5	202.9	6	177	6	159.3	6	151	8	150	8
United States	232.8	3	231.3	3	229.5	4	208.7	4	218.2	4	236	4
Australia	258.3	2	250.1	2	268.1	2	274.0	2	279.5	2	290.5	2
China	371.0	1	413	1	38.2	1	478.2	1	453.5	1	450.1	1
Russia	211.9	4	230.1	4	248.8	3	247.5	3	249.5	3	253.5	3
Indonesia	120.1	7	89	10	109.6	9	116.4	9	176.3	5	168.2	5
Peru	187.6	6	185	5	187.7	5	173.0	5	175.9	6	164.5	7
Canada	107.7	8	108.2	7	133.3	7	152.1	7	159	7	165	6
Uzbekistan	71.4	11	73.3	11	77.4	12	81.4	11	83.2	11	82.9	12
Ghana	91.0	9	95.8	8	107.4	10	107.4	10	95.1	10	95	10
Papua New Guinea	63.5	13	56.5	13	60.5	13	56.3	14	57.2	14	59.9	13
Mali	43.5	17	43.5	15	48.2	16	47.4	16	49.0	15	49.8	15
Brazil	67.3	12	67.3	12	80.1	11	81.2	12	81.8	12	83.3	11
Tanzania	49.6	15	49.1	16	46.6	17	45.8	17	46.8	18	48.7	16
Chile	44.5	16	48.6	17	48.6	15	44.2	18	n/a	n/a	n/a	n/a
Philippines	37.1	19	41	18	40.5	20	42.8	20	46.7	19	48.5	17
Argentina	59.3	14	54.6	14	50.1	14	59.7	13	63.8	13	57.4	14

Mexico	88.6	10	95.3	9	119.8	8	117.8	8	135.8	9	120.5	9
Colombia	37.5	18	39.1	20	41.2	19	43.1	19	47.6	17	48.3	18
Zimbabwe	n/a	n/a	n/a	n/a	0	21	n/a	n/a	n/a	n/a	n/a	n/a
Kyrgyzstan	n/a	n/a	n/a	n/a	0	0	n/a	n/a	n/a	n/a	n/a	n/a
Venezuela	n/a	n/a	n/a	n/a	0	0	n/a	n/a	n/a	n/a	n/a	n/a
Kazakhstan	36.7	20	40	19	42.6	18	48.9	15	48.2	16	48.0	19
Dep. Rep. of Congo									45.7	20	44.4	20
Rest of the World	452.9	-	465	-	506.4	-	550.6	-	548.5	-	554.3	-
World Total	2,838.1	-	2,860.6	-	3,061.5	-	3,131.5	-	3208.6	-	3222.3	-

*Revised by GFMS

Source: Gold Fields Mineral Survey (GFMS) 2017

The country maintained its position as the tenth leading producer of gold in 2016

6

1.4 History of Mining in Ghana

Mining is defined as the activities relating to the extraction of "any substance in solid or liquid form occurring naturally in or on the earth, or on or under the seabed, formed by or subject to geological process including building and industrial minerals but does not include petroleum or water" (Ghana's Minerals and Mining Law, PNDCL 153). Gold is by far the largest and most important mineral resource in Ghana in terms of production and contribution to government revenues and employment, making up 93% of all mining exports and attracting 60% of all foreign investment (MC, 2004). The country is known to have one of the world's largest gold ore reserves, and its production ranked as 10th in the world (Table 1.1) or second in Africa in 2016 (Chamber of Mines, 2016). The country hosts the second-largest gold deposits in the Africa region after South Africa. Consequently, the nation derives a bulk of its external revenues from gold mining forming as much as 90% of the total mineral exports of Ghana. Currently, ten (10) large-scale mining companies are producing gold. There are also one (1) large scale company each producing bauxite, manganese, and diamond. There are also, over 650 registered small scale mining groups are engaged in the mining of gold, diamonds and industrial minerals, in addition to ninety (90) mine support service companies (Chamber of Mines, 2004). Ghana is also the third leading African producer of manganese ore and a significant producer of bauxite and diamonds, although their contribution to the national economy and government revenues is much smaller.

Despite its rich mineral endowment, Ghana's mining sector stagnated for four decades before the 1980s because of economic, financial, institutional, and legal problems that impeded investment in the sector. However, the mining sector has seen an impressive growth since the launch of the Economic Recovery Program (ERP) in 1983. The program attracted US$4 billion of private investment in the first five years after its inception. The World Bank saw the reforming of the mining sector as a means to alleviate the economic crisis that befell on the country in the 1980s. During that period the government of Ghana attracted over $4 million in foreign investment through privatisation of its large-scale mining sector for the development and expansion of large-scale gold mining and explorations activities alone (Hilson and Potter, 2003). From that time to date mining has overtaken cocoa as the biggest single foreign currency earner. According to Aryee (2000), the reform of the mining sector, therefore, has produced a dramatic boom in investment flows which have made the national economy quite dependent on the sector.

Currently, mineral exploration is continuing at more than 200 sites, and the government has approved 58 mining leases for gold and other minerals; the total land area is approximately 3,000 km2. Ten large-scale and one medium-scale gold mining companies are currently operational. The growth rate of minerals production, however, has slowed down and turned to negative over the last few years. Several others face capacity constraints and are estimated to have a remaining lifetime of only a few years without heavy investment. Figures from the Minerals Commission indicate that the mining industry attracted an amount of US$ 1,000 million in total investment inflow into the country in 2012. These investments came from producing, exploration and support service companies. The multiplying effect of these investments in Ghana's economy cannot be overestimated.

Among the new entrants in gold mining is Newmont, the world's largest gold producer, which is developing two projects in Ghana: Ahafo and Akyem, with combined golf reserves estimated at 16 million oz. The government also plans to boost production of other minerals, notably bauxite, with a view to develop an integrated aluminium industry in the country. Talks are under way with Alcoa, world's largest mining producer of aluminium, for this venture. A key constraint in further expansion of industrial-scale mining is the lack of an essential infrastructure in terms of adequate power, water, and transportation. The Ghana Railway Company, the only bulk mineral carrier in Ghana, is at present in a difficult financial situation, which will need to be addressed to ensure the smooth shipment of ore from the mines to the Takoradi Port.

Table 1.2 Major mineral production in Ghana (1990-2016)

Year	Gold (Ounces)	Diamond (Carats)	Bauxite (M/t)	Manganese (M/t)
1990	522,517	484,877	381,373	364,373
1991	946,269	702,172	352,921	325,964
1992	1,006,943	596,236	338,244	353,476
1993	1,251,010	584,848	423,747	294,789
1994	1,396,887	746,949	426,128	271,989
1995	1,630,309	627,319	512,977	245,432
1996	1,550,814	714,717	473,218	161,690

1997	1,644,622	698,585	504,401	273,224
1998	2,353,000	823,125	442,514	348,406
1999	2,257,681	680,343	355,260	638,937
2002	2,315,000	627,000	503,825	638,937
2001	2,205,473	870,490	715,455	1,212,338
2002	2,115,196	924,638	647,231	1,132,000
2003	2,208,154	927,000	494,716	1,509,432
2004	1,794,497	911,809	498,060	1,593,778
2005	2,149,372	1,062,930	726,608	1,714,797
2006	2,244,680	970,751	885,770	1,658,701
2007	2,486,821	837,586	748,232	1,156,339
2008	2,585,993	598,042	693,991	1,089,021
2009	2,930,328	354,443	490,367	1,012,941
2010	2,970,080	308,679	512,208	1,194,074
2011	2,924,385	283,369	400,069	1,827,692
2012	3,166,483	1,490,634	752,771	1,490,634
2013	3,192,648	2,003,176	826,994	2,003,176
2014	3,167,755	241,120	798,114	1,353,486
2015	2,848, 574	174,188	1,014,605	1,288,624
2016	4,131,440	141,530	1,143,676	2,018,254

Source: 1   1970 -1999 - Minerals Commission, Statistical overview Of Ghana's mining industry
2   2000 – 2013 Chamber of Mines Annual Reports
3   2014-2016 Minerals Commission

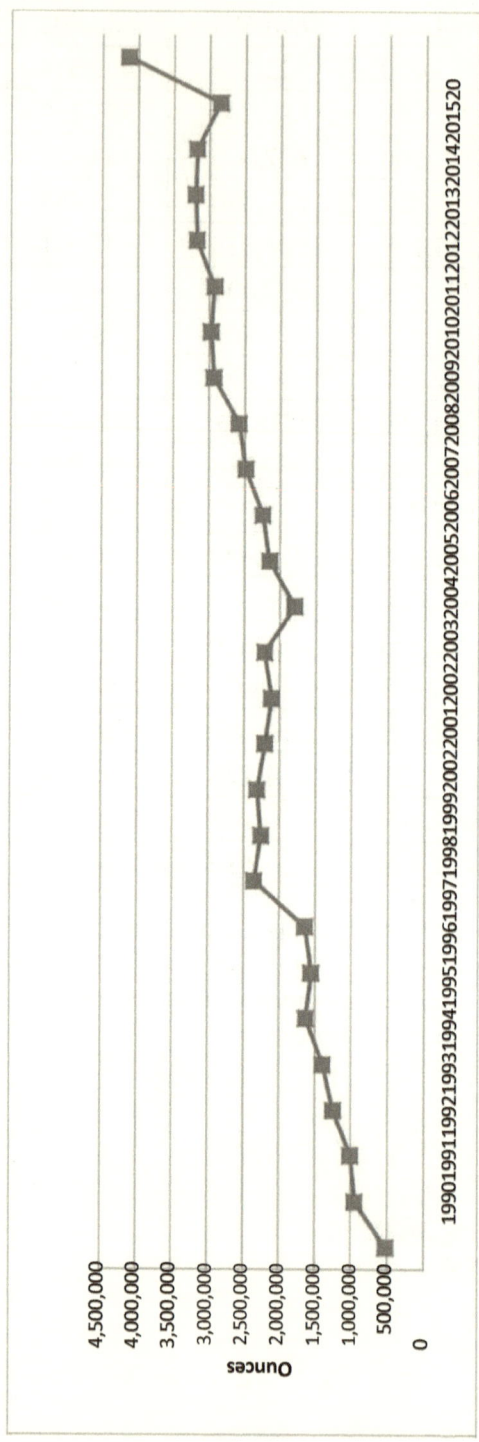

Source: Ghana Chamber of Mines Annual Reports
Figure 1.1 Production trend for Gold (Only Producing Member Companies of the Chamber

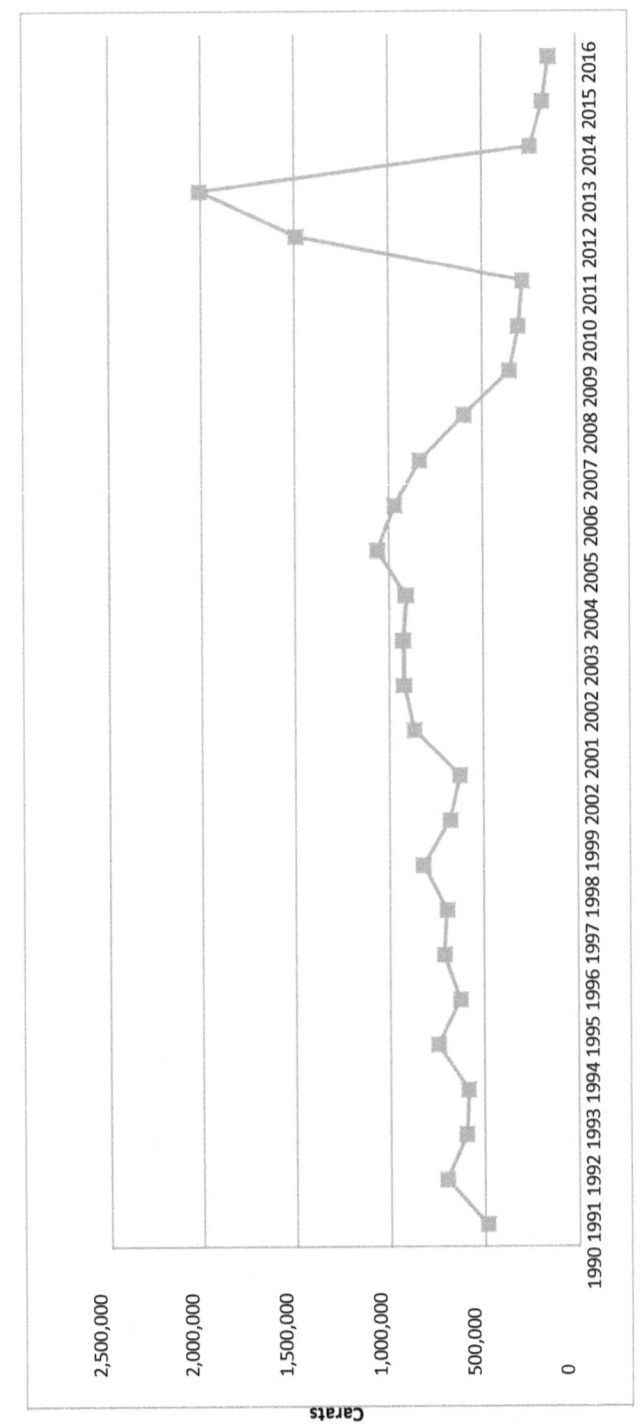

Figure 1.2 Export trend for Diamond

Source: Ghana Chamber of Mines Annual Reports (2016)

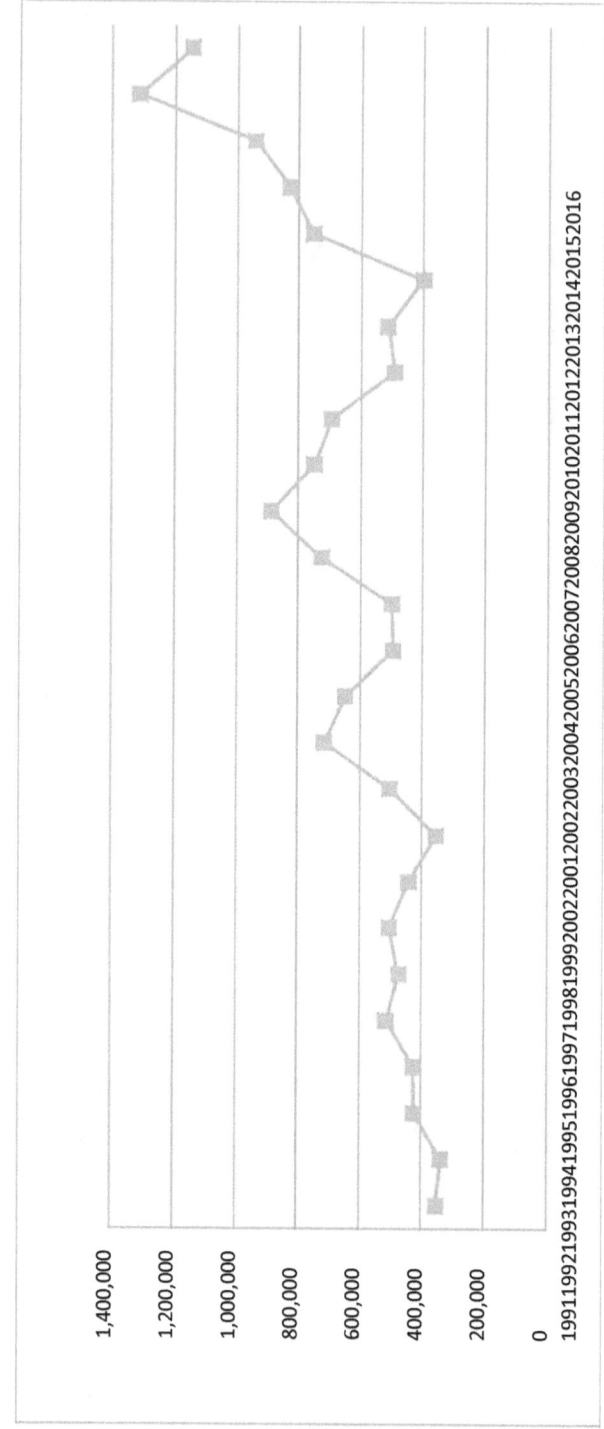

Exports of Bauxite (1990–2016)

Source: 1990-2013 Ghana Chamber of Mines Annual Reports 2 2014-2015 Minerals Commission 3 2016 Bank of Ghana
Figure 1.3 Export trend for Bauxite

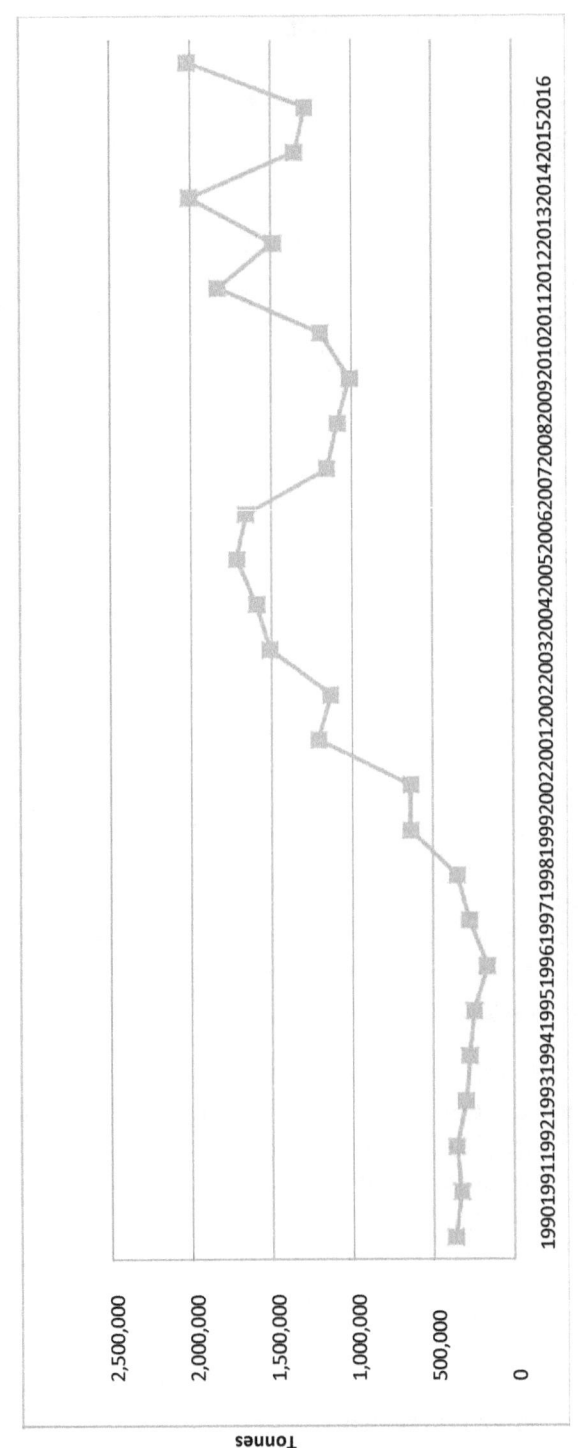

Exports of Manganese (1990-2016)

Source: Ghana Chamber of Mines Annual Reports (2016)

Figure 1.4 Export trend for Manganese

1.5 Socio-economic Contribution of Mining in Ghana

Mining is considered as one of the important economic activity in Ghana due to the significant contribution to the country's export earnings, government revenues, and employment. Mineral revenue represented about 4.1% in the national GDP and about 9% of government revenues, and the formal mining sector employed some 15,000 workers in 2004 (Minerals Commission, 2004). The Minerals Commission estimates that ASM, often called "galamsey," might account for an additional 500,000 people. Many of those involved in ASM are women and children, and a significant number of them are informal participants in the sector.

1.5.1 CSR awareness in Ghana

Over the past years, MCGs have applied portions of their profits to support national or local projects. This so-called "corporate social responsibility" constitutes an expression of intent; no explicit legal requirements exist for mining companies to provide services to local communities. Indications are that in the major extraction sites, mining companies account for a significant share of the social infrastructure and provision of social services. These typically involve educational and health care services, infrastructure and utilities, and the sponsoring of recreation and sports activities. During a 2004 conference on Corporate Social Responsibility in Ghana, the Chamber of Mines stressed the significant fiscal contributions by the mining companies at the local level. The implementation of the Extractive Industries Transparency Initiative will clarify the level of contribution the companies provide. Improved CSR trends within the sector have arisen out of pressures from civil society and NGOs, rather than from the compliance efforts that the EPA has been able to exert. Violence in the sector is not unknown; for example, reports persist of security agents shooting people who trespass in mining areas. A number of civil society organizations are advocating improved environmental governance, social protection, compensation for affected communities, and human rights in mining regions in Ghana.

1.5.2 Revenue, Taxation and Royalties

Revenue from gold mining also dropped in 2004 despite a steady price increase of gold. This could be attributed mainly to decreasing production capacity in several large mines such as Obuasi and Bogoso. Total production of manganese, bauxite, and diamonds in 2004 amounted to US$30.2 million, US$10.6 million, and US$26 million, respectively (CM, 2004). Gold exports totalled US$2.25 billion in 2008, up from US$1.3 billion in the previous year. Total merchandise exports from the mining sector amounted to US$2.35 billion in 2008. The mining industry's revenues contribute significantly in insulating the country against the adverse impacts of the global economic downturn. According to the Bank of Ghana, total merchandise export earnings by the traditional minerals (Gold, Bauxite, Manganese and Diamond) represented about 49% of gross merchandise exports. The mining sub-sector grew remarkably by 11.2% compared to the 6.8% it recorded in 2009. By this growth performance, the industry came second behind the Electricity sub-sector which grew by 16.7% in 2010.

The total mineral revenue of MCGs rose significantly from US$2,925,831,036 in 2009 to US$3,724,847,388 in 2010 representing an increase of 27% mainly on account of the healthy price of gold. Out of this Gold revenue went up by 27% from US$2,842,821,528 in 2009 to US$3,620,766,467 in 2010. The industry committed about 68% of its revenue into the Ghanaian economy with an average of 25% going through the Bank of Ghana to satisfy its statutory obligations and a further 43% through the commercial banks. These inflows to the economy provided a significant cushion to the local currency as it performed quite creditably against the country's trading currencies. Furthermore, as have been agreed by the MCGs, one percent of mineral revenue was paid as dividends to other shareholders and voluntarily contributed an amount of US$ 17.6 million to its host communities and general public.

Data from Ghana Chamber of Mines (2016) state that producing mining companies made capital expenditure of about US$ 452.9 million on plant, machinery and equipment in 2016. This accounted for 13.9 per cent of total mineral revenue. According to data from the Central Bank, the minerals industry consolidated its position as the country's leading export earner by improving its share in gross merchandize exports from 32.2 per cent in 2015 to 45.5 per cent in 2016. Cocoa and crude oil followed with respective shares of 22.3 per cent and 12.5 per cent. Thus, the proceeds from export of minerals is a little more than twice that of cocoa and more than three times the out turn of crude oil in 2016. These can be depicted in Table 1.3 below.

The domestic economy of Ghana is known to revolve around subsistence agriculture, which accounts for 50% of GDP and employs 85% of the workforce. In 2009 the public sector wage increased as a result of the introduction of the Single Spine Pay Policy and this has led to continued inflationary deficit financing and depreciation of the Ghanaian currency (the Cedi). Although the economy have been considered as challenged, Ghana remains one of the more economically sound in many countries in Africa, with twice the per capita output of the poorer countries in West Africa. The mining sector performed well by contributing significantly of 6.8% to GDP. In Table 1.3 below the sector contribution to exports rose from 36.26% in year 2000 to 45% in 2008. This was considered to be the single largest contributor from 1991, with the exception of 2004 when it was overtaken by the cocoa sector.

The provisions of 1987 Minerals (Royalties) Regulations guides the payment of royalties by mining companies. The payments are to be made quarterly and are based on the profitability of mining operations. The royalty rate ranges from a minimum of 3% to a maximum of 12%, but in reality, it has been fixed for the minimum 3% in past years regardless of the gold price. But the Minerals and Mining Act, 2006 fixed the royalty rate ceiling at 6% based upon the estimated operating profit margin calculated for a mine prior to starting production. The royalty payments are said to be accounting for an average of about 98% of the total royalties paid to government over the last 10 years.

Table 1.3 Major mineral's Revenue in Ghana (2004-2016)

Year	Gold (USD)	Diamond (USD)	Bauxite (USD)	Manganese (USD)
2004	731,232,992	26,010,921	10,615,787	30,247,600
2005	903,899,619	34,729,560	18,022,283	39,028,514
2006	1,327,455,774	30,021,594	22,564,545	39,932,287
2007	1,711,511,381	26,366,914	19,686,731	36,831,651
2008	2,202,878,021	18,892,292	19,810,287	62,348,266
2009	2,842,821,528	6,991,088	11,157,480	64,860,940
2010	3,620,766,467	11,126,089	15,145,755	77,809,077
2011	4,630,255,619	14,850,558	13,406,433	119,989,551
2012	5,309,042,207	11,163,107	28,495,592	98,605,517
2013	4,610,284,057	8,030,808	32,923,689	135,475,951
2014	3,841,579,039	10,700,962	36,258,720	91,147,458
2015	3,320,635,208	6,424,888.81	41,063,160	70,581,339
2016	5,152,782,855	4,305,484	36,865,766	104,807,091

Source: 2004-2013, Chamber of Mines Annual Reports
2 2014-2016 Bauxite data- Bank of Ghana

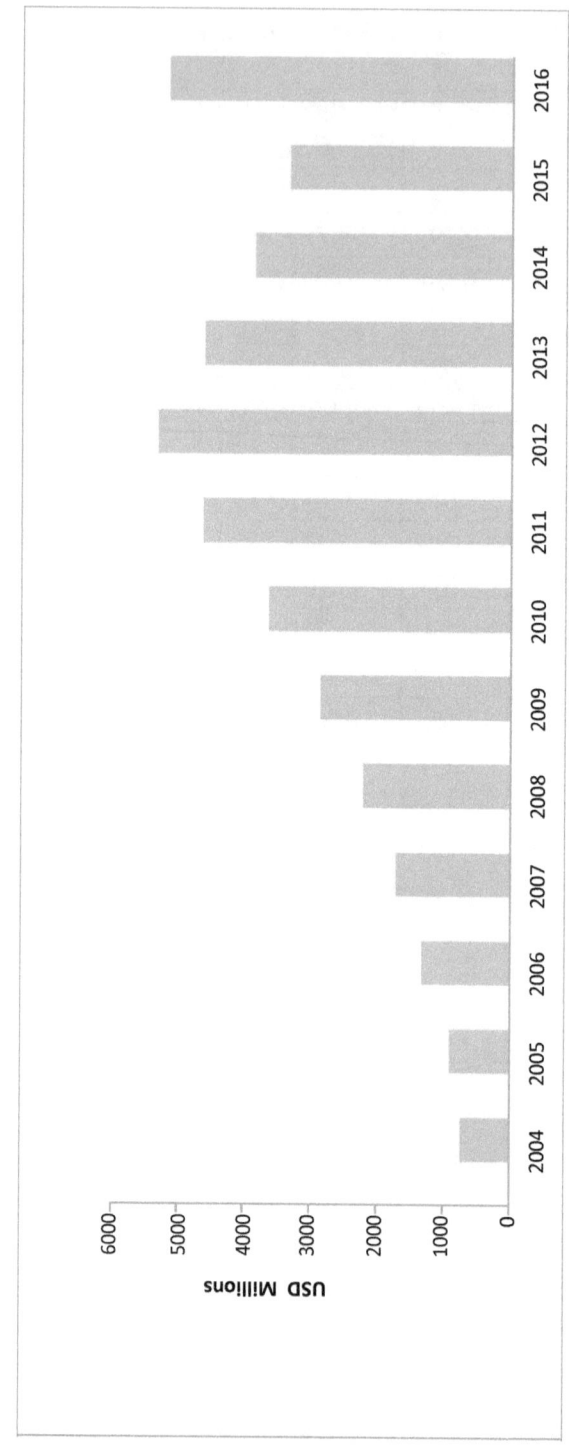

Source: Ghana Chamber of Mines Annual Reports
Figure 1.5 Revenue trend for Gold (US Dollars)

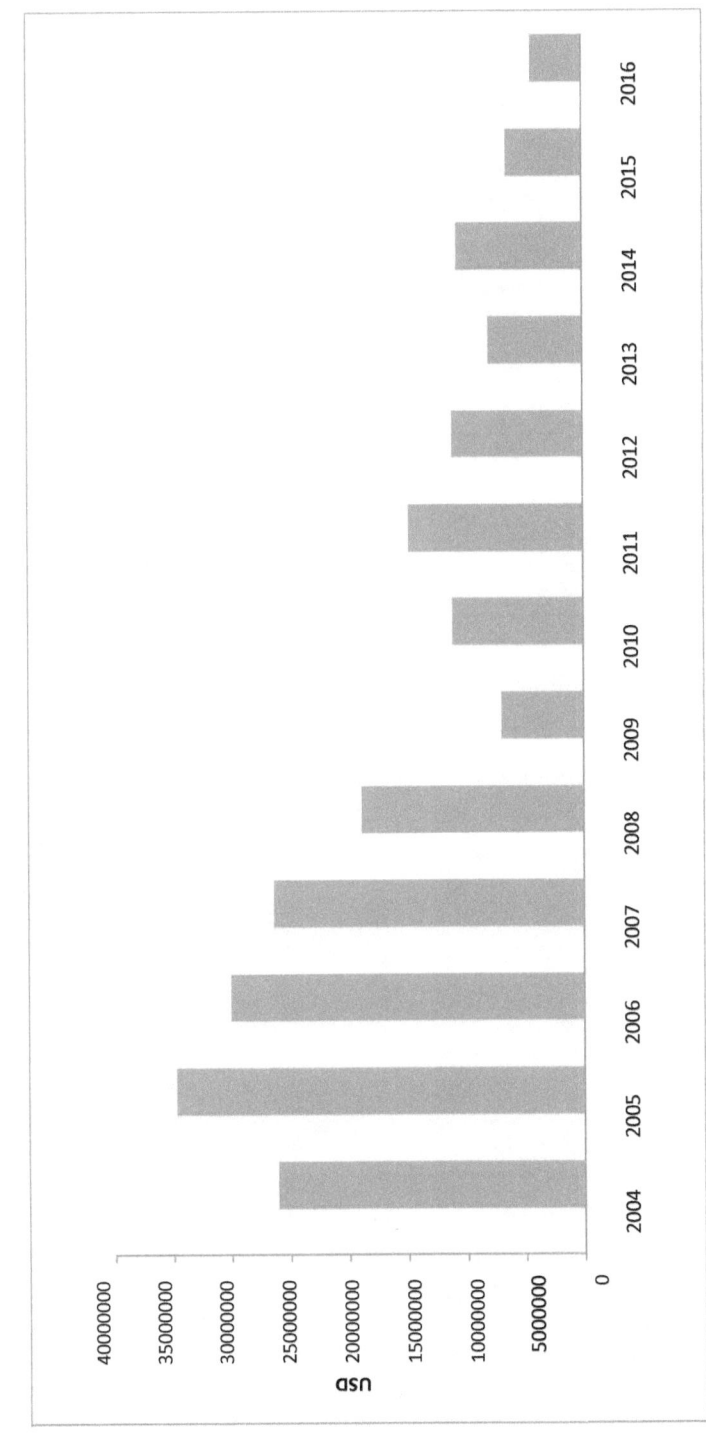

Source: Ghana Chamber of Mines Annual Reports
Figure 1.6 Revenue trend for Diamond (US Dollars)

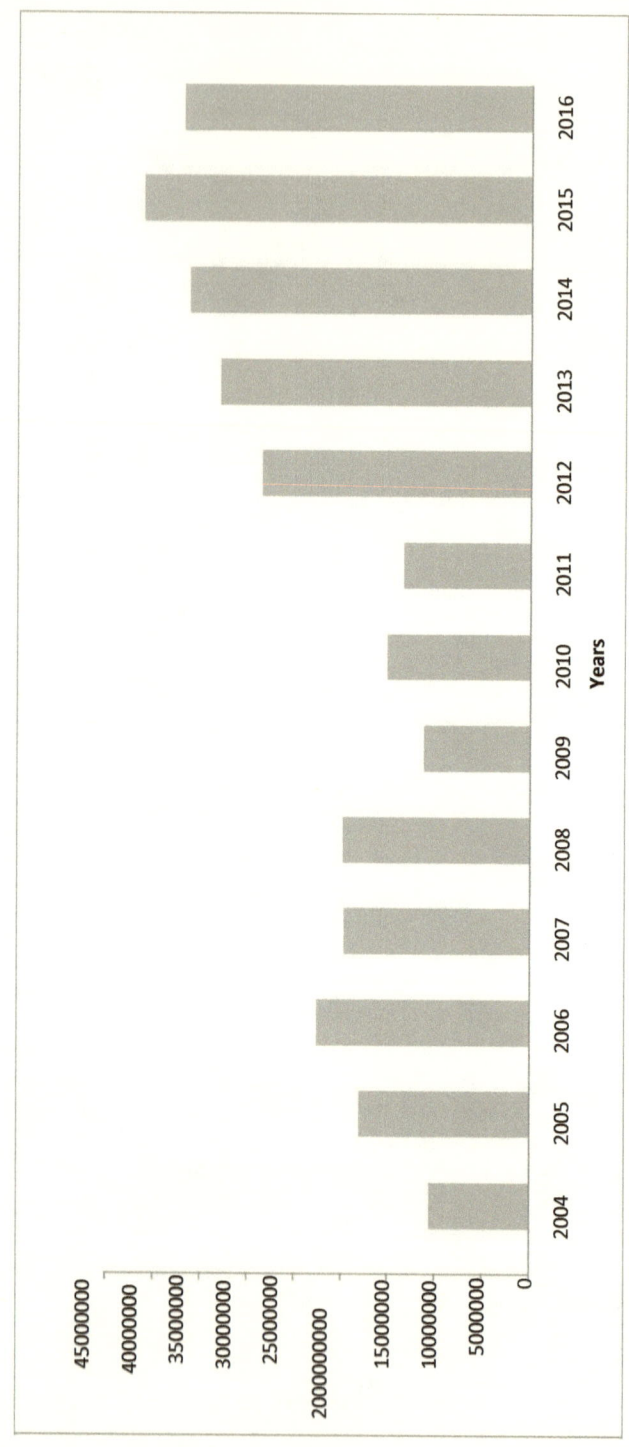

Figure 1.7 Revenue trend for Bauxite (US Dollars)

Source: Ghana Chamber of Mines Annual Reports

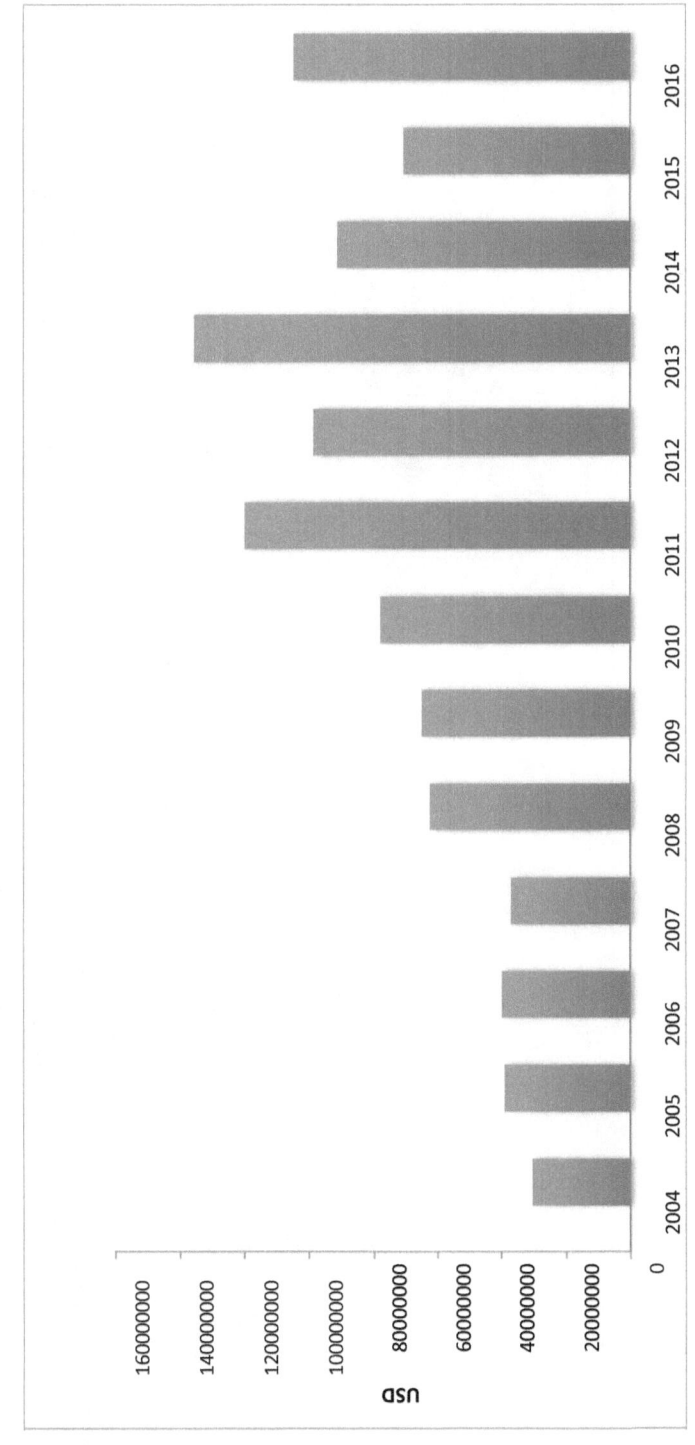

Table 1.4 Mining sector Contributions (1990-2016)

Year	Corporate Tax	Mineral Royalties	PAYE	Reconstruction Levy	Total Mining Contribution To GRA	Total IRS (GRA)	%Mining To Total
1990	2,825,941,158	1,893,436,000			4,719,377,158	52,818,068,300	8.94%
1991	821,844,979	3,021,277,000			3,843,121,979	61,485,625,496	6.25%
1992	455,051,883	4,545,804,000			5,000,855,883	74,931,531,366	6.67%
1993	4,393,447,293	7,485,121,000	2,649,306,000		14,527,874,293	113,236,997,000	12.83%
1994	7,214,082,000	12,783,689,000	4,810,802,000		24,808,573,000	166,595,941,000	14.89%
1995	20,392,973,000	20,911,926,000	7,951,763,000		49,256,662,000	275,513,201,000	17.88%
1996	9,160,528,000	35,527,027,000	16,834,543,000		61,522,098,000	424,491,908,000	14.49%
1997	9,868,796,000	34,594,950,000	25,022,023,000		69,485,769,000	605,782,577,000	11.47%
1998	14,450,773,000	49,841,242,000	31,016,506,000		95,308,521,000	785,436,693,000	12.13%
1999	31,117,108,000	48,620,419,161	27,839,260,000		107,576,787,161	901,663,758,000	11.93%
2000	15,789,167,000	118,736,935,173	59,243,800,000		193,769,902,173	1,409,445,273,000	13.75%
2001	24,812,893,000	127,358,386,430	76,111,678,000	4,251,467,579	232,534,425,009	1,950,162,751,000	11.92%
2002	23,501,158,000	153,452,471,032	101,457,668,000	26,474,633,878	304,885,930,910	2,757,747,781,032	11.06%
2003	68,137,702,000	194,387,579,429	141,049,450,000	16,785,882,702	420,360,614,131	3,824,078,389,429	10.99%
2004	100,331,114,000	215,743,706,000	134,357,711,000	36,346,622,100	486,779,183,100	5,333,114,704,000	9.13%
2005	269,889,639,000	235,951,903,000	194,058,939,000	22,957,004,700	722,857,485,700	6,446,385,048,000	11.21%

2006	404,361,775,000	316,254,789,000	216,525,776,000	11,085,262,400	748,227,602,400	7,333,916,866,000	10.20%
2007*	47,415,690	40,882,042	34,587,597	–	123,021,866	901,242,340	14.42%
2008*	73,554,697	59,004,892	47,139,242	–	179,978,383	1,222,272,177	15.32%
2009*	124,600,880	90,415,902	103,061,985	–	319,022,676	1,731,633,034	18.21%
2010*	241,578,780.28	144,697,000	132,469,709.91	–	519,682,174.41	2,441,331,841.81	21.29%
2011*	649,902,536	222,024,706	161,822,107	–	1,034,221,712	3,746,024,194	27.61%
2012*	893,773,828	359,392,853	207,495,934	–	1,461,202,977	7,461,202,977	27.04%
2013*	518,545,259	364,673,038	220,131,571	–	1,104,047,315	5,900,057,003	18.71%
2014*	441,235,058.84	470,356,948.81	259,459,815.44	–	1,172,117,330	7,622,600,239.22	15.38%
2015*	463,128,598.35	137,708,716.67	404,743,477.20	–	1,354,379,971.47	9,091,499,085.61	
2016	696,978,327.50	550,738,649.73	399,925,811.85	–	1,648,183,048.75	10,425,117,616.68	

Source: Minerals Commission, Statistical overview of Ghana's Mining Industry (1990–2003); Internal Revenue Service/ GRA (2004–2016)

* Contributions for 2007 to 2016 are in GH¢. All others are in Cedis, (¢10,000: GH¢1)

Table 1.5 Mining's Contribution to total Royalties (1993-2016)

Year	Mining Mineral Royalties ('000)	Total Royalties ('000)	% Mining Contribution
1993	7,485,121		98.8%
1994	12,783,689		99.2%
1995	20,911,926		98.8%
1996	35,527,027		97.2%
1997	34,594,950		99.1%
1998	49,841,242		97.3%
1999	48,620,419		97.4%
2000	118,736,935		98.6%
2001	127,358,386		99.0%
2002	153,452,471		99.5%
2003	200,867,945		99.3%
2004	215,743,706		98.2%
2005	235,951,903		96.3%
2006	316,254,789		88.5%
2007*	40,882		97.7%
2008*	59,005	62,915	93.8%
2009*	90,416	96,653	93.5%
2010*	144,697	150,539	96.1%
2011*	222,024.7	224,966.1	98.7%
2012*	359,392.9	363,805.6	98.8%
2013*	364,673.04	656,248.6	55.6%
2014*	470,356,948.81	1,061,350,968.10	44.32%
2015*	485,632,656.88	1,128,321,098.88	44.04%
2016*	550,738,649.73	–	

Data source: Minerals Commission, Statistical overview of Ghana's Mining Industry (1993-2003) Internal Revenue Service/GRA (2004-2015)
* Contributions for 2007 to 2015 are in GH¢. All others are in cedis, (¢10,000: GH¢1)

1.5.3 Employment

Mining sector has also contributed immensely in the area of employment. The sector as at the end of the year 2010 provided 15,861 out of which 2% were expatriates with the rest being Ghanaians. Employee remuneration, which covered salaries and other benefits accounted for 7% of mineral revenue. Out of this, ninety-eight percent (98%) were Ghanaians and 2% expatriate. Small-scale mines also generated about 500,000 jobs across the nation and certain indirect jobs were also created, because of the existence of the mining companies. According to Ghana Chamber of Mines (2016) at the end of 2016, total direct employment by the producing member companies stood at 11,628, representing a 16 per cent increase relative to the outturn of 9,939 in 2015 (refer to Figure 1.9).

The 2016 employment level comprises 11,438 Ghanaian employees and 190 expatriates, with the latter representing 1.63 per cent of employees. Compensation, wages and salaries to these personnel amounted to US$ 435.64million in 2016. Producing mining companies of the Ghana Chamber of Mines paid an amount of GH¢ 77,261,440 as social security contributions on behalf of their employees in 2016. These are long-term funds, which assist the country's capital formation drive. Indeed the mining industry contributes to capital formation multipliers, which arise from mining companies' influence in attracting foreign capital to the country and domestic capital formation.

Source: Ghana Chamber of Mines
Figure 1.9 Direct Employment in the Mining Sector (1994–2016)

1.5.4 Voluntary contributions to socio-economic by mining sector

One of the factor which makes mining important in the country is the contribution it makes to social multipliers, which arise from the role of mining companies in the development of human resources and infrastructure such as schools, colleges, clinics, roads, and housing. These are done in relation to their CSR practices where all producing members of the Chamber of Mines have set up social development funds to make available such facilities to their communities. All the gold producing members of MCGs have decided to contribute a dollar per ounce as well as a percentage of gross profit into the funds for the socio-economic development of host communities in which the mine is cited. There are also situations where some companies provide lump sums annually for the development of their communities in addition to this. Besides its notable contributions to the economy, the mining industry contributes to social multipliers which arise from the role of mining companies in the development of human resources and infrastructure such as schools, colleges, clinics, roads, and housing. Specifically, Table 1.5 below shows the contributions made by the MCGs to host communities and the general public within the period of 2008 to 2016 in the areas of Education, Health, Electricity, Roads, Water, Housing, Agro-Industry, Agriculture, Sanitation, Resettlement Action Plan, Alternative Livelihood and Projects (others). In 2016, the producing members of the Chamber invested US$ 12.29 million in their host communities though a reduction from US$ 17.10 million from the previous year.

Table 1.6 Socio-Economic Contributions

Socio-economic contributions	2007 (US$)	2008 (US$)	2009 (US$)	2010 (US$)	2011 (US$)	2012 (US$)	2013 (US$)	2014 (US$)	2015 (US$)	2016 (US$)
Education	1,010,246	1,406,203	1,259,262	2,826,680	1,767,790	2,627,407	1,615,766	1,958,895	4,881,435.26	1,295,078
Health	565,596	415,711	777,486	1,055,260	407,052	696,267	1,479,904	2,278,188	1,245,160.87	1,185,726
Electricity	458,797	333,611	285,319	526,218	1,917,227	675,837	405,801	601,149	368,239.52	396,225
Roads	609,146	2,612,992	1,375,626	1,459,049	1,368,497	1,619,277	2,029,584	1,464,607	1,166,873.44	3,295,488
Water	220,876	649,703	284,668	678,976	1,437,529	1,001,818	1,561,361	724,822	677,889.62	343,497
Housing	618,531	686,523	112,635	155,266	165,633	277,236	264,198	522,284	75,357.93	508,729
Agro-Industry	386,668	778,970	50,624	-	54,982	155,544	130,561	57,004	-	-

Agriculture	743,937	1,551,893	610,353	809,228	1,407,674	1,531,716	374,280	354,172	83,685.72	54,097
Sanitation	262,863	405,236	227,844	196,428	202,762	404,767	583,384	689,408	242,125.46	208,555
Resettlement Action Plan	4,503,381	567,820	800,188	1,190,371	29,345,274	7,583,140	68,624	3,025,322	4,814,535.77	3,380
Alternative Livelihood										
Projects (others)	2,897,767	992,973	798,658	2,214,584	175,935	1,853,298	1,236,646	4,295,824	960,471.14	489,250
Others	830,747	2,004,228	2,841,649	6,478,411	5,528,313	8,250,046	2,373,943	4,890,261	3,312,122.35	4,508,936
Total	13,108,554	12,405,861	9,424,309	17,590,469	43,732,833	26,676,354	12,124,052	20,861,936	17,827,897.09	12,288,960.86

Source: Ghana Chamber of Mines Annual Reports

1.6 Summary

This Chapter presented provided information about the country Ghana in relation to the overview and historical review of the mining operations in the country in Sections 1.2 and 1.3. Section 1.4 presented the policy and legislative framework governing the mining operation including Mining rights and payment of Royalties. This followed by Section 1.5 which presented the importance of the mining and CSR awareness in Ghana for example revenue from Taxes.

CHAPTER 2

Regulatory and Policy Framework

2.1 Introduction

The chapter also considered the Mining Institutions and Regulatory and Policy Framework which have been established to control the mining operations in Ghana. Such regulatory framework are Minerals and Mining Act 2006 (Act 703), The Minerals Commission, Environmental Protection Agency, Ghana Chamber of Mines, The Forestry Commission Act, 1999 (Act 571) and the Water Resources Commission Act, 1996 (Act 522) respectively. The legislative framework for the mining industry in Ghana is regarded as comprehensive and thorough although there is no specific guideline or legislation on CSR in the country (SRC Consult, 2010). The following paragraphs explain some of the importance of the regulatory framework of the sector.

2.2 Legal Systems

Prior research on corporate disclosure suggests that common law countries generally have legal systems that are strong enough to protect investors (Ball et al., 2003; La Porta et al., 1998). Legal systems, therefore, shape the orientation of companies and accounting regulation. Consequently, common law gives priority to individual rights as opposed to the state, because of its decentralized nature. Common law systems assure that transactions are done between or among different independent parties, hence, they lead to a high level of disclosure to the public (see Beck et al., 2003; Ho and Wong 2001).

The scope of service provided by professional accountants is influenced by legislation and case law (Bushman and Piotroski, 2006). Investors expect that their investments will be protected against fraud and losses (Mensah et al., 2003). They also expect that if their rights are violated they can seek satisfactory redress from the system. The protection of investors is therefore enshrined in the legal system that is efficient, effective and swift. Such a system is very significant in attracting investors and in providing capital for the development of a country. Ghana is a common law country and its legal system builds on the foundation of the British common law tradition supplemented by customary (written and unwritten) laws. Lopez-de-Silanes (2003) suggest that countries with a common law legal framework are oriented toward quality disclosure. It can be argued that since Ghana applies common law, it will have a better accounting and disclosure environment than code law countries within Africa.

2.3 Companies' Code

The companies' code, known as The Code, prescribes the basic reporting requirements by companies in Ghana but stops short of suggesting applicable accounting standards (Assenso-Okofo et al., 2011). Since Ghana was ruled by the United Kingdom (UK), the Companies' Code 1963 Act 179, which is the corporate legal framework of Ghana, has its roots in English Common Law. It was written by the same person called D. Gower, who also is also known to have written the UK Companies Act. In addition to the adopted international codes and standards, The Code presents guidelines for the governance of all companies incorporated in Ghana as well as recognizing the role of auditors in disclosure and transparency in operations of corporate bodies (Assenso-Okofo et al., 2011).

The regulatory framework governing the commercial aspects of auditing and financial reporting in Ghana emanates from the Companies' Code, Act 179, Accounting Standards, and regulations promulgated by the SEC and the GSE. All entities, both public and private companies, incorporated under the law of Ghana must comply with these requirements.

Public companies are required to prepare quarterly financial statements and yearly reports and accounts and disseminate the audited accounts to their members (IFAC, 2006). There is no clear guide in the Companies' Code as to which standards to follow in the preparation and audit of annual financial statements (ICAG, 2000). Private companies are to comply with the Companies' Code disclosure requirements (e.g. presenting audited accounts to every member of the company) and Ghana National Accounting Standards.

One of the limitations of the Code is that preparation of a cash flow statement or statement of changes in equity was not mentioned as part of the financial statements and income statement were referred to as profit and loss account (Assenso-Okofo et al., 2011). Furthermore, the Code has been seen virtually no amendments since its promulgation in 1963 and this may encumber accounting reporting and disclosure practices (Assenso-Okofo et al., 2011). According to Assenso-Okofo et al. (2011) Companies' Code of Ghana is seen as outdated in the light of current accounting and disclosure practices and calls for revision to adequately protect investors and to be in tune with current practices in the world of accounting. There is also the need for the Code to take into the some form of regulations for the sustainability reporting practices in Ghana.

2.4 Policy and legislative framework

Ghana's long-term policy objective in the minerals sector has been guided by the need for establishing a legal and macroeconomic environment that would attract investments in new exploration and encourage the expansion of existing mines. This policy objective has led to the evolution of a supporting legislative and institutional framework that encompasses the Constitution, various laws, regulations, and instruments that provide guidance on how mining operations ought to be conducted. Recently passed, the Minerals and Mining Act 30 is the primary law governing the sector. The legal framework draws a clear distinction between large scale mining operations and artisanal and small-scale mining (ASM) operations. In fact, the laws and regulations provide generous incentives and benefits for large-scale mining operations but have failed to create a similar incentive and benefit structure for ASM.

2.5 The Minerals and Mining Act 2006 (Act 703)

The law on mining in Ghana is laid down in the Minerals and Mining Law, 1986, PNDCL 153 (Law 153) as amended by the Minerals and Mining Amendment Act 1993, Act 475 (Act 475). With reference to this legal framework, the State owns all the minerals in their natural state within Ghana's land and sea territory, including its exclusive economic zone. This means that all minerals in Ghana are vested in the President by the power of this framework, on behalf of and in trust for the people of Ghana. Mineral Legislation Between 1986 and 2006, the Mineral and Mining Law 1986, PNDCL.153 was the basic mining legislation but was amended to take into consideration the changes in the international mining. The current Minerals and Mining Act, Act 703 of 2006 became the law for Ghana's minerals and mining sector after a protracted review from the early 2000s.

2.5.1 Mining rights

There are two types of procedures currently exist for granting and keeping mining rights. One procedure relates to ASM, and the second procedure pertains to other mining operations that are not considered as small scale. Both procedures require a license in order to exercise a mineral right. Although the law does not explicitly state which licenses are applicable to ASM, by virtue of the capital intensity involved with these licensees, they appear to exclude ASM. Of interest is the fact that the license period varies considerably depending on the type of mining operation. In the case of large-scale mining operations, an application for the grant of a mineral right must provide a statement specifying: (1) particulars of the financial as well as technical resources available to the applicant for the mineral operations; (2) an estimated amount of the operations' expenditure; (3) particulars of the proposed mineral operations, and (4) particulars of the applicant's proposal with respect to employing and training Ghanaians.

When the application is submitted to the Minerals Commission it is expected that within 90 days of receipt of the application, the MC must submit its recommendations on the application to the minister. The small-scale mining license for is issued by the Minister for Mines or someone authorized by him and is only issued to Ghanaian nationals aged 18 years or older who are registered by the District Office in the respective area. Once licensed, a person is authorized to "win, mine, and produce minerals by an effective and efficient method and shall observe good practices, health and safety rules and pay due regard to the protection of the environment during mining operations. Licenses granted to both large-scale and small-scale miners are not transferable except with the minister's prior approval. By comparison, small-scale miners have no such requirement; however, this license may only be transferred to a Ghanaian citizen. In addition, the holder of a mineral right is required to pay a prescribed annual ground rent as well as an annual mineral right fee.

According to Minerals and Mining Act a mining right may be suspended or cancelled where the holder of the right: (1) fails to make payments on the due date required by the law; (2) becomes insolvent, bankrupt, enters into any agreement or scheme of composition with his creditors, or takes advantage of an enactment for the benefit of his debtors or goes into liquidation, except as part of a scheme for an arrangement of amalgamation; (3) makes a statement to the minister in connection with the mineral right that he knows or ought to have known is false; (4) or for any reason becomes ineligible to apply for a mineral right under the Minerals and Mining Act. One interesting fact is that the law does not specify breaches of law as grounds for the suspension or cancellation of mining rights. Before the suspension or cancellation the mineral right, the minister gives notice to the mineral right holder as prescribed by law.

In addition to environmental impacts, mining operations frequently pose social concerns. For example, large-scale mining operations often cause involuntary resettlement, resulting in loss of land, livelihoods, and resources for local communities. Land and resource rights of indigenous communities in and around mining concessions are often a bone of contention. As surface mining operations become more widespread in Ghana, land-use conflicts are bound to escalate-surface mining usually requires that huge swaths of land be cleared of vegetation, and top soil and soil nutrients are lost. Villagers have often been dispossessed of their farmlands to make way for mining, resulting in loss of livelihoods and traditional community values and linkages. ASM miners believe they have rights on surface and subsurface land which is contrary to the law that vests mineral rights with the state. Local chiefs, who are authorities at the local level, often endorse this position. Altogether, local people call into question the existing rights and procedures for miners to access land. These rights are perceived as unfair and biased against landowners.

2.6 The Forestry Commission Act, 1999 (Act 571).

The Forestry Commission (FC) was re-established under the Forestry Commission Act, 1999 (Act 571) to regulate the utilisation of forest and wildlife resources, the conservation and management of those resources and the coordination of policies related to them. Before a mining operation can be made the company must obtain a permit from the FC according to mining, Section 18 of Act 703. After securing the permit, the holder is monitored by a committee comprising the FC, Ministry of Lands and Natural Resources, Minerals Commission, Environmental Protection Agency and the District Assembly. A regular report is required from the holder by the committee and any holder who operates outside this framework could lose any mineral rights they have and also be sanctioned appropriately.

2.7 The Water Resources Commission Act, 1996 (Act 522)

The Water Resources Commission (WRC) is another legislative regulator which was established under the Water Resources Commission Act, 1996 (Act 522). The WRC has the main responsible of regulating and managing the utilisation of water resources. The organisation has the function of granting the water rights and the co-ordination of any policy in relation to them. Under Section 17 of Act 703, a holder of a mineral right may, for purposes of the mineral operations may obtain, divert, impound, convey and use water from a river, stream, underground reservoir or watercourse within the land where the mineral right have been granted subject to obtaining the requisite approvals or licences under Act 522. The Commission is equipped with the power to enter upon any land to inspect construction works to ascertain the amount of water abstracted or capable of being abstracted as a result of the works being carried on and prescribe sanctions for breaches.

2.8 Summary

The chapter considered the Mining Institutions and Regulatory and Policy Framework which have been established to control the mining operations in Ghana. Such regulatory framework. The Minerals and

Mining Act 2006 (Act 703), The Minerals Commission, Environmental Protection Agency, Ghana Chamber of Mines, The Forestry Commission Act, 1999 (Act 571) and the Water Resources Commission Act, 1996 (Act 522) respectively were also discussed in this chapter.

CHAPTER 3

Mining and Accounting Institutions

3.1 Introduction

Many researchers have argued that an efficient and effective institutional framework and a favourable socio-economic and political climate can improve the accounting and reporting practices of a country (see Ali and Ahmed, 2007; Ball, et al. 1999; Williams and Tower, 1998: Assenso-Okofo et al., 2011). According to Assenso-Okofo et al. (2011) accounting is the product of its environment. A favourable environment is vital for capital providers, both domestic and international, to supply funds for businesses to continue and thrive. In general, emerging countries suffer from a dearth of investments due to lack of accountability caused by lack of information on how the managers have used their resources but it would be too simplistic to conclude that this is true of all emerging countries since each country has its unique environments that must be understood (Assenso-Okofo et al., 2011).

Although accounting academics and practitioners have focused on understanding reporting environments in many emerging nations (see Ashraf and Ghani, 2005 on Pakistan; Ali and Ahmed, 2007; Mashayekhi and Mashayekh, 2008; Al-Akra, et al., 2009), there are few studies carried out in African countries. There is limited information on accounting practices during the pre-colonial period in Ghana as to when and how accounting practices in Ghana began (Wilks, 1989). The establishment of the Institute of Chartered Accountants of Ghana in 1963 by Act 170 of Parliament is seen as the foundation of formal accounting in Ghana. The main legal framework for financial reporting and auditing for both private and public companies is based on the Companies' Code of 1963 (Act 179). In addition to the Companies' Code, which governs corporate financial reporting practices, other accounting regulations include the Ghana Accounting Standards (until 2009), Securities Laws, Taxation Laws, and Corporate Governance Guidelines. In this section, the author takes a look at some of the socio-economic and political environment, the institutional framework, accounting education, the role of local and international accounting standards, and the existing enforcement mechanisms in order to provide an understanding of the sustainability reporting by mining companies in Ghana.

3.2 The Minerals Commission

The Minerals Commission (MC) is the institution mandated to take charge of the regulation, utilization and coordination of the mineral resources and implementation of related policies. The main duty of MC is to foster the development of mineral resources by attracting foreign investors and negotiating leases with them. The Commission is therefore serves as the technical advisory agency to Government. In addition to these functions the Inspectorate Division (DI) of the Commission is given mandated to enforce the mining regulations. The ID of the Minerals Commission was established under Section 101 of Act 703 and it is responsible for enforcing the Mining Regulations, 1970 (L.I. 665) or its amendments which ensures health and safety in mining operations in Ghana. The ID is expected to be satisfied with a proposed mining project and issues before an operating permit is granted. The head of the ID, the Chief Inspector of Mines, is mandated under Act 703 to inspect all aspects of any mining operations for compliance, including whether the nuisance is being created handling to ensure that the proposed mineral operations would be or is being carried out safely.

The commission also has the responsibility of promoting the formalization of Artisanal and Small-scale Mining (ASM) but does not consider environmental or social aspects of regulation within its mandate. The MC's regulatory role effectively ends once licenses are issued. A sound coordination between the MC and EPA is crucial to ensure that environmental and social issues are integrated into the process of granting mining licenses. Interestingly, the MC cites the EPA as the major source of delay in the process of granting mining exploration and exploitation licenses. There are also several occasions that the MC and EPA have also been sued for their failure to execute statutory responsibility to ensure compliance when the company reneged on its duty to restore the environment.

3.3 Environmental Protection Agency

The Environmental Protection Agency (EPA) was established under the Environmental Protection Agency Act, 1994 (Act 490) with the main aim to enforce and monitor the environmental regulations in Ghana. The Mission and vision of the EPA is to co-manage, protect and enhance the country's environment in particular as well as seek common solutions to global environmental problems in accordance with Section 18 of Act 703 and the Environmental Assessment Regulations, 1999 (L.I. 1652). A holder of a mineral right requires an environmental permit from the EPA in order to undertake any mineral operations. The main legal framework used by the EPA for regulating and monitoring mineral operations is the Environmental Assessment Regulations, 1999 (L.I. 1652). The EPA achieve its aim through an integrated environmental planning and management system established on a broad base of public participation, efficient implementation of appropriate programmes and technical services, giving advice on environmental management as well as effective and consistent enforcement of environmental laws and regulations. It is an implementing agency, a regulatory body and catalyst for change towards sound environmental stewardship and are mandated to prevent, reduce, and as far as possible eliminate toxic pollution and actions that lower the quality of life.

Over the past two decades, major players in the mining industry have increasingly recognized both the need and obligation to identify and mitigate the adverse environmental consequences of their activities. For example cyanide spillages. Cyanide is commonly used for the recovery of gold from the ore in industrial-scale, hard-rock gold mining in Ghana. Other problems with mining include dust and noise pollution from blasting, and risk of water-borne diseases such as malaria from water collected in mine pits. In recent times, most large-scale mining operators have elaborate and comprehensive environmental impact assessment and management plans as well as having the commitment to sustainable. The regulations state that an environmental impact assessment (EIA) is mandatory for the mining and processing of minerals in areas where the mining lease covers a total area of more than 10 hectares.

A holder of a mineral right is required to submit an annual environmental report in respect of the mineral operations to the Agency. Mineral right holders based on approved work plan for reclamation. The EPA undertakes monitoring activities regularly to ensure that mineral right holders are compliant with the terms of the environmental permit and the environmental laws generally. With respect to sanctions, the EPA is empowered to suspend, cancel or revoke an environmental permit or certificate and/or even prosecute offenders when there is a breach. Although Ghana has made important strides toward the development of a comprehensive legal and regulatory framework governing the mining sector, many important legal factors continue to impede progress. For example, issues of enforcement and compliance with mining laws and regulations continue to be a major challenge. The EPA, which is tasked with the responsibility of monitoring undertakings relating to mining, has neither the financial nor human capacity to effectively monitor all undertakings. Other legal constraints include ambiguities in the law that leave room for wide interpretations as well as omissions in some instances.

3.4 Ghana Chamber of Mines

It was in 1903 when the West Africa Chamber of Mines, were set up with the principal objective of advancing and protecting the mining interests of the shareholders. The Chamber was composed of directors of the Mining Companies in London who, as one of its functions, had power to promote or oppose any legislative measures or petition government and administrative bodies on many matters which directly affected mining interests in the colony. On 6th June 1928, the Gold Coast Chamber of Mines was incorporated as a private company and operated at Tarkwa in the Western Region. This was further changed to the Ghana Chamber of Mines (GCM) when Ghana's attained independence on 6th March 1957. On 6th May 1960, the form of the objects of the chamber was also altered by a special resolution and on 14th February 1964, the chamber was incorporated under the Companies Code 1963 (Act 179) as company Limited by Guarantee. All the programmes and activities of the Chamber are funded entirely by its members.

The vision of GCM is to be a respected, effective and unified voice for the mining industry. With its Mission Statement being to represent the Mining Industry in Ghana using the resources and capabilities of its members to deliver services that address members, government and community needs in order to enhance development. The principles that will guide decision making (core values) which the members of the Chamber will not compromise whilst achieving the mission and pursuing the vision are: Honesty, Transparency, Good Governance, Good Corporate Citizenship, Commitment and Unity. Some of its main objectives are to (1) promote and protect the interests of the MCGs; (2) promote and protect the image of the mining industry; (3) establish and maintain effective membership governance; and (4) provide thought leadership for the solution of national issues related to mining.

The Chamber was established as a voluntary private sector employers' association representing companies and organizations engaged in the minerals and mining industry in Ghana. Programmes and activities of the Chamber are funded entirely by its Member Companies, which are largely responsible for producing almost all of Ghana's minerals. As a voluntary association of public and private sector actors of the mining industry, promotes CSR and human resource development within the industry, and is supporting the Extractive Industries Transparency Initiative (EITI). The MCGs often are said to be competing with ASM for limited mining areas, but their strong bargaining powers, plus the government's desire to attract foreign investment, often result in marginalization of ASM. The GCM operates through an extensive committee system, which enables the specialist expertise and the intellectual capital within the MCGs to be tapped in a collective effort to enhance the overall business environment in which the mining industry and the country can have the opportunity to thrive and flourish.

There are five categories of membership, namely, Represented, Pre-production, Contract Mining, Exploration, and Affiliate categories. The Represented are further divided into two namely Level A and Level B. Level A consist of ten (10) companies which are the MCGs in commercial production. Level B is only one (1) company which actually use the ore to manufacture its final product. Pre-production which also consists of four (4) members, are those about to go into commercial production. They become Represented Members after being in commercial production for one year. The Contract Mining is a class of membership those providing contract mining services. They are also consisting of four (4) members. Exploration also has six (6) memberships for prospecting/exploration MCGs. Finally, Affiliate have twenty (20) memberships is engaging in mining and minerals related services mostly referred them as Services Industries. There are a total of forty four (44) member companies of the MCG and other affiliate members.

The Code of the Ghana Chamber of Mines requires the signatories to uphold fundamental human rights, to respect the culture and customs of their employees and local communities affected by their actions. It includes provisions on good governance and consultation with local communities. Most MCGs are signatories to this Code of Conduct. Again, the Chamber has developed a Sustainable Alternative Livelihood Policy (SALP) which focused on creating long-term employment opportunities, primarily in the mining districts, beyond direct employment provided by the MCGs which is consistent with the provisions of Section 10 of its Code of Conduct. Table 3.2 shows the principles of SALP.

Table 3.1 Principles on Sustainable Alternative Livelihood Programmes (SALP)

1. Consult their host communities on their aspirations, and values regarding development and operation of mineral projects, recognizing that there are links between environmental, economic, social and cultural issues.
2. Voluntarily contribute to the socio-economic development of their host communities as far as their resources will allow
3. Promote transparency and active participation of local communities and stakeholders in all aspects of the Sustainable Alternative Livelihood Programmes (SALP), including planning, implementation and monitoring.
4. Promote accountability through formal meetings (i.e., Annual General Meetings -AGMs) and public documents to review strategies and progress in achieving the defined outcomes.

5. Establish an SALP Coordinating Committee in the communities where they work with representation from the communities, chiefs, opinion leaders and local political authorities.
6. Establish sustainable and adequate funding for SALP that ensures that cyclical global metal prices do not adversely affect member companies' ability to fund projects during downturns. In this light, member companies shall set up funding mechanisms with a clearly defined source and mode including effective and efficient fund management committees.
7. Promote projects that achieve long-term sustainability and community acceptance and ownership.
8. Communities participating in the program will be required to contribute either financial or in-kind to achieve the sustainable outcome.
9. Collaborate with institutions and agencies to provide skills, entrepreneurship and business skills development and training to assist them to be employable.
10. Where possible, develop cooperatives based on existing governance structures (e.g. Farmers' Association) and facilitate the registration of beneficiaries as co-operatives through registration at the Registrar Generals Department. This enhances the ability of the beneficiaries to meet orders of their clients.
11. Encourage entrepreneurs in the communities and the country as a whole to participate in identified opportunities within the value chain of the mining company's operation.

Source: Ghana Chamber of Mines

3.5 Institute of Chartered Accountants, Ghana (ICAG)

The Institute of Chartered Accountants, Ghana (ICAG) was established by an Act of Parliament to regulate the accounting profession. The main mandate of ICAG is to ensure that professionals acquire the appropriate education and practical experience to qualify as a practicing accountant or auditor (Assenso-Okofo et al., 2011). The ICAG is the sole body responsible for regulating and conducting exams although other consulting firms help students prepare for professional examinations. To be qualified as an accountant in Ghana, an individual must pass the ICAG exams or obtain an equivalent foreign qualification from either the Association of Chartered Certified Accountants (ACCA), the American Institute of Certified Public Accountants (AICPA), the Institute of Chartered Accountants of Scotland (ICAS), or the Institute of Chartered Accountants in England and Wales (ICAEW) (Assenso-Okofo et al., 2011).

According to Mantey (2007) lack of centrally coordinated curriculum and a strong teaching methodology as well as weak critical thinking skills by students brought about the establishment of Institute of Professional Studies (IPS), a private initiative that was later adopted by the government to provide quality accounting tuition for students (Assenso-Okofo et al., 2011). Its curriculum emphasizes teaching, learning and research opportunities for students to develop analytical skills in problem solving (Assenso-Okofo et al., 2011) and this helped the students to acquire professional accounting, auditing skills, and practical experiences (IPS, 2005). This is expected to improve the reputation of the ICAG and put it in a position to champion good financial reporting and disclosure (ROSC, 2004). To a large extent, the quality of financial reporting depends on the strength, size, competence, adequacy and effectiveness of the accounting profession (Ali and Ahmed, 2007). It has been argued that a well-developed accounting profession leads to improved, judgmentally based public accounting systems rather than centralized and uniform systems (Radebaugh and Gray, 1993).

The ICAG, which regulates accountancy profession in Ghana, is the sole institution with the right to award Chartered Accountant designation to qualified members. Therefore ICAG members are recognized under the Companies' Code to engage in audits of companies' accounts in Ghana. The ICAG also expects its members to follow the code of professional conduct, which is based on the abridged version of the IFAC code of ethics, and has the power to discipline those who violate the code. The ICAG is governed by an eleven-member council, four of which are appointed by the sector minister (Ministry of Education-Government). Despite this authority, the ICAG faces constraints including limited financial resources, operational difficulties, a lack of adequate capacity to monitor, investigate and discipline members who do not comply with the existing standards and code of ethics.

3.6 The Ghana Stock Exchange (GSE)

The Ghana Stock Exchange (GSE) was incorporated in July 1989 and started trading in November 1990 and is governed by a council. The number of listed companies has grown from 11 in 1990 to 38 in 2015 (Ghana Stock Exchange Fact book, 2015). The slow in growth were as a result of lack of understanding of the stock exchange, limited knowledge and low purchasing power among citizens as a result of high inflation and high interest rates (Assenso-Okofo et al., 2011). According to GES, the main criteria for list include capital adequacy, profitability, spread of shares, years of existence and management efficiency The existence of GSE has also has seen significant capital appreciation and index performance with an average rate of capital appreciation of 70% since 1990 apart from 1995 and 1996, which saw 6.3% and 20% appreciation respectively (Ghana Stock Exchange Fact Book, 2007). There are three main methods of gaining entry to GSE namely introduction, private placement and offer to the public. The GSE's capitalization of GH¢11.47 billion (US $2380 million) as of June 30, 2007 (Standard and Poor's, 2008) makes it one of the largest in sub-Sahara Africa in terms of market capitalization (The World Fact book, 2007).

According to Assenso-Okofo et al. (2011), Ghana financial market is dominated by the money markets, even though it is made up of bonds, equities, and foreign exchange and derivative markets (2011). Despite the potential and recent achievements, the GSE has a number of challenges which calls for reforms. There appears to be weak institutional foundation, capacity issues and enforcement gaps that need to be addressed to achieve the required standard (Senbet and Otchere, 2006). However, ROSC (2005) is of the view that improvement in Ghana's capital markets will depend very much on the institutional capacity of the regulators, administration and judiciary rather than on reforming the legal framework. An efficient and vibrant stock market will most likely lead to quality accounting reporting and disclosure practices (Assenso-Okofo et al., 2011) and can increase investors' confidence that will attract capital as well as promotion of good accounting and disclosure practices (Adhikari and Tondkar, 1992; Gray, et al., 1990; Assenso-Okofo et al., 2011). This is due to the fact that apart from monitoring and enforcement by sanctioning non-compliant companies, stock markets compel companies to make available to the public their annual reports and financial reports (Camfferman and Cooke, 2002).

Ghana instituted the Securities Industry Law 1993 (PNDCL 333) and established the Securities and Exchange Commission (SEC) through the advice of the World Bank to formulate how the SEC should function. Some of the functions of the SEC are to maintain surveillance over the securities market, to ensure adequate protection and security to all investors and to guarantee fairness in the securities market (Assenso-Okofo et al., 2011). It is also the responsibility of SEC provide for the establishment of stock markets, the licensing of stockbrokers/dealers and investment advisors, oversees issues concerning accounting and auditing, and conduct of securities business. It also investigates any breaches and settles disputes arising under the securities laws and the Companies' Code. In December 2000, the Securities Industry Law 1993 (PNDCL 333) was amended significantly (Securities Industry Law 2000 (Act 590) to provide for the operation and regulation of Unit Trusts and Mutual Funds to eliminate insider trading abuse, reviewed takeovers, mergers and acquisition of companies (Assenso-Okofo et al., 2011).

3.7 International Financial Reporting Standard (IFRS) Adoption in Ghana

Many developing countries adopted IFRS in full or in part in reporting so as to be accepted in the international community and to prevent the problems that arise where there is a limited resources in terms of human, technical, logistics or otherwise to prepare national standards (Ashraf and Ghani, 2005; Ball et al., 2003). Until 2009, the accounting practices in Ghana followed Ghana Accounting Standard (GAS), which were meant to reflect International Accounting Standard (IAS). However, GAS differed significantly from IAS. It is also observed that the ICA of Ghana and the GNASB are neither addressing revision of the standards nor bringing the procedures and processes in line with international standards. In the light of this disparity and in order to eliminate the gap that exists between the national and the international standards, Ghana has moved away from just adaptation to the adoption of IFRS. A process started from January 2007 from public companies, banks and insurance companies, then from January 2009 extended to small-and medium-sized private enterprises as well as government ministries, departments and agencies in the public sector (Assenso-Okofo et al., 2011: Khalid et al., 2013).

In spite of the adoption of the IFRS, a good financial reporting or disclosure is still considered as a challenge. The technical and logistical capacity to review financial statements and to detect accounting/auditing violations is lacking even at the Registrar-General Department. Even though Registrar-General of companies in Ghana has the legal authority to sanction companies for non-compliance, but the timely filing of AR of non-listed public and private companies is not enforced. Furthermore SEC/GSE who have been mandated to monitor the listed companies for IFRS compliance and timely submission of reports lack the capacity and resources to effectively undertake their mandated work (Assenso-Okofo et al., 2011). The ICAG has no effective and efficient mechanism to ensure compliance with auditing standards and professional code of ethics, therefore faces difficulties and challenges in monitoring and enforcing compliance of accounting and auditing standards. ICAG has thus been tasked to facilitate continuous education for its members, teachers, analysts and the general public on the effects and application of IFRS.

One study asserted that mining companies generally report on their global operations and that in some cases such companies disclosure site-specific information on social and environmental issues and has increased quantity, quality and complexity of environmental and social disclosure (Jenkins and Yakovleva, 2006). Furthermore, Jenkins and Yakovleva (2006) found that companies are increasingly disclosing information on their websites with site-specific reports and updated news items relating to social, environmental, community and employee issues on their websites.

3.7.1.1 IFRS 8 Geographic Segmental Reporting

Segmental and geographic reporting IFRS 8 which came into force on 1 January 2009 covers segmental reporting. Operating segments are identified based on internal reporting of financial information to the board which can be geographical or segmental based on products and services. According to Tonkin and Skerrat (1989) segmental reporting can be described as the reverse side of the coin of consolidation (Khalid et al., 2013). As organizations become larger and more complex, academics, regulators, analysts and other users have identified a need for disaggregated information (Edwards and Smith, 1996 p.156). Even though majority of the disclosures in IFRS 8 are numerical, there are some narrative disclosures: factors used to identify the operating/reportable segments, the nature and effect of changes from the prior period in the methods used to determine segment results, and the nature and effect of any asymmetrical allocations between segments. It can be seen here that there is no specific requirement from IFRS 8 to report narrative social and environmental information by segment or geographic sector.

From the perspective of accountability, transparency and governance, segmental reporting can be useful to a diverse range of stakeholders, not only to shareholders but also to other user groups such as local communities and indigenous populations. Segmental reporting is considered important due to varying rates of profitability, opportunities for growth and risks between sectors and geographical locations (Haller and Park, 1994, p.563). Indeed, the usefulness of segmental reporting has been emphases by researchers, for example, Kochanek (1974) stated "theoretically, segmented income statement and balance sheet data should enable the analyst or investor to precisely analyze component parts of the firm and thereby evaluate the firm's stock on a more rational basis" (Khalid et al., 2013). Segmental reporting is also known to provide relevant information to the financial market and especially to financial analysts (Boersema and Van Weelden, 1992; Deppe and Omer, 2000).

Segmental reporting is expensive and complex to produce due to technical issues, and cost has been a matter of academic debate for many decades (Baker and McFarland, 1968; Mautz, 1968; Boersema and Van Weelden, 1992; Khalid et al., 2013). These additional costs required can serve as a deterrent to produce what is effectively a voluntary environment of reporting. Another concern raised was that technical issues related to segment reporting are reflected in higher audit fees, since it requires extra work from auditors (Sanders et al., 1999). Furthermore, the use of such information could cause potential disadvantages due to competitors and other parties having access to the information (Khalid et al., 2013). These can reveal to competitors weaknesses or opportunities to be exploited to their own advantage (AICPA, 1994; Sanders et al., 1999; Deppe and Omer, 2000).

3.8 Summary

This chapter presented provided information about the mining and accounting institutions in Ghana. The main functions of Minerals Commission and Environmental Agency were discussed. The mandate of Ghana Comber of Mines being an organisation of commercial mining companies in Ghana and the relation with mining institutions were also presented in this chapter. Finally Institute of Chartered Accountant and Ghana Stock Exchange function as far as mining is concerned were also presented.

CHAPTER 4

Global Reporting Initiative Guidelines (version 3.0)

4.1 Introduction

Global Reporting Initiative (GRI), is one of the frameworks of SR which has assumed more popularity, most prominent and most widely adopted a number of companies preparing SR (e.g., Morhardt et al., 2002; Joseph, 2012), give meaning to the application of accounting to sustainability (Joseph, 2012). The framework came about as a result of the need for a consistent CSR reporting standard. GRI was founded in 1997 by the Coalition for Environmentally Responsible Economies (CERES) and the United Nations Environmental Programme (UNEP) and was initially published in 2000 (Isaksson and Steimle, 2009). The information guiding the preparation of SR can be obtained from the GRI website, particularly the "Sustainability Reporting Guidelines" Version 3.0 (or G3 Guidelines) and CorporateRegister.com. The GRI is a network-based organisation that pioneered the world's most widely used sustainability reporting framework. It is a long-term, multi-stakeholder, international process whose mission is to develop and disseminate globally applicable SR guidelines. GRI, as compare to other guidelines, provides detailed guidelines on "how to report," defining overall goal and content using principles and guidance, and "what to report" or determining content using standard disclosures and sector supplements (Joseph, 2012). The main focus of the analysis is on the key documented guidelines on concepts, measures and assurance, which was "developed through a unique multi-stakeholder consultative process involving representatives from reporting organizations and report information users from around the world" (GRI, 2010, p. 44).

The GRI (2010) defines SR as the practice of "measuring, disclosing, and being accountable to internal and external stakeholders for organizational performance towards the goal of sustainable development" (GRI, 2010, p. 3). This goal of sustainable development is used to mean, "meeting the needs of the present without compromising the ability of future generations to meet their own needs." The GRI guidelines specifically have the main principle to achieve transparency. This transparency has been defined as the complete disclosure of information on the topics and indicators required to reflect impacts and enable stakeholders to make decisions (Joseph, 2012). The GRI concepts appeared to highlight transparency and sustainable development, with inter-generation equity forming the preferred theme underlying sustainability. However, GRI did not specify the contents of the report, but allowed for variations in the elements to comply with the initiatives. GRI also includes some general recommendations on what the report should contain, such as the extent to which the report preparer has applied the GRI Reporting Framework (including the Reporting Principles) in the course of reaching its conclusions. While the total firms reporting on sustainability and the nature of the reports continue to evolve, the statistics serve to provide an indication of the diversity of SR trends. Thus, GRI is representative of the SR, having a wide global application of firms from a variety of industries and of global (multi-national) nature.

GRI became independent in 2002, and published its second Sustainability Reporting Guidelines that year as the foundation upon which all other GRI reporting documents are based. The framework provides common guideline for sustainability reporting worldwide and is made to complement and strengthen financial reporting to shareholders. These guidelines are for voluntary use by organisations for reporting on the economic, environmental, and social dimensions of their activities, products, and services (GRI, 2010) to a broad and diverse range of stakeholders including the communities. The GRI aims to develop a globally applicable framework for an organisation's Sustainability and CSR reports. There are over 900 companies spread throughout 50 countries currently report on the basis of the GRI guidelines (Roca and Searcy, 2012).

The GRI is probably one of the most important initiatives as supplementary guidelines relevant to mining and metal sector has been developed to provide common framework guidelines for mining and metal companies. Their main purpose is to support companies in preparing SRs that integrate social, environmental and economic impacts of business on their stakeholders. The GRI is an internationally accepted framework that promotes comparable sustainability reporting (Isaksson and Steimle, 2009). The current version of the guidelines (GRI-G3) was published in 2010 which contains principles and guidance for defining content and quality of the sustainability report as well as for setting the report boundaries. In order to meet the continuous improvement and application worldwide GRI is currently in the process of developing G4, its fourth generation of sustainability reporting guidelines. An exposure draft of G4 was released for public consultation from 25 June to 25 September 2012 and can be still be accessed from the GRI webpage. GRI's core goals include the mainstreaming of disclosure on environmental, social and governance performance.

Many research has suggested that sustainable development indicators such as jobs, water usage, pollutant emissions, solid wastes, rehabilitation and land use, energy source and consumption, and health and safety are relevant for mining industry reporting (Byrne et al., 2002; Hilson and Basu, 2003; Azapagic, 2004; van Berkel and Bossilkov, 2004). The used of GRI framework by many companies was as a result of the many advantages associated with it; (1) the GRI framework is considered to have international acceptance; (2) has a rigorous framework for the application of triple bottom line reporting; (3) it was put together by a wide variety of experts after Stakeholder consultation; (4) the GRI guidelines are easily accessible and (5) it is made to fit for all types of companies (Willis, 2003; Lamberton, 2005; Reynolds and Yuthas, 2008; Farneti and Guthrie, 2009). As its main purpose is to support companies in the preparation of SRs that integrate social, environmental and economic impacts of business, the GRI intends to establish guidelines that promote comparable SR (GRI, 2010). The framework contains principles and guidance for defining content and quality of the SR. It also has principles for setting the report boundaries, i.e. the determinants the parameters of the report. These principles and guidance that are expected to be followed in the preparation of SR are explained below.

4.2 Principles of Reporting

The G3 guidelines provide principles that could provide guidance in determining what the stakeholders want. Several concepts were found in the guidelines but not in a cohesive form. The sustainability context, for example, contained distinctions between the global and local view about sustainability reporting, indicating the underlying principles of SR and how organizations contribute to sustainability trends at the local, regional and global contexts (G3 guidelines, 2010).

Reporting principles define the processes to be followed in the preparation of the SR such as selection of the topics and indicators as well as how to report on them. The principles consist of a definition, an explanation, and provide a medium to assess its use by the reporting company (GRI, 2010). The provision of medium of assessment are intended to provide some tools for self-diagnosis, but not as specific disclosures to report against as well as serving as a reference for explaining decisions about the application of the Principles. According to the GRI (2010) to ensure a balanced and reasonable presentation of the performance of the company, the content the report should cover must be determined by considering the purpose and experience, expectations and interests of the company and its stakeholders. By helping to decide what to include in the report the reporting principles enable companies in achieving transparency in SR preparation (GRI, 2010). The Principles are divided into two namely Principles for Defining Report Content and Principles for Defining Report Quality.

4.3 Principles for defining report content (PDRC)

The PDRC as the name is concerned, defined the process to be followed to identify what should be disclosed in the SR. The principles take into consideration the activities, impacts, and the substantive expectations and interests of both the companies and its stakeholders (GRI, 2010) when preparing the SR. These principles are designed to be used to define the content of SR are explained in the following paragraphs.

4.3.1 Materiality

Organizations are encountered with various ranges of topics and areas to report and materiality is one of the principles that companies must consider when deciding the type of content to be included in the SR (GRI, 2010). The SR is expected to cover the company's significant economic, environmental and social impacts on its stakeholders. The topic must be relevant to the extent of substantively influence the assessments and decisions of stakeholders. A topic of area is considered relevant those items may reasonably be considered important for reflecting the organization's economic, environmental and social impacts, and, therefore, potentially merit inclusion in the SR (GRI, 2010). The principle of materiality is the threshold at which aspects become sufficiently important that they should be included in the SR.

4.3.2 Stakeholder Inclusiveness

The SR is made not for its own sake; it is prepared to satisfy its stakeholders. This means that the identification by the company of its stakeholders and their inclusion is one of the principles in defining the SR content (GRI, 2010). It will also help in explaining how it has responded to their reasonable expectations and interests. The term stakeholders here are used to mean investors, suppliers, creditors as well as those who have other relationships to the organization. The reasonable expectations and interests of stakeholders must be taken into consideration in the preparation of the SR.

4.3.3 Sustainability Context

With the sustainability context principle the SR should be prepared to disclose the organization's performance in the wider context of sustainability information on performance. According to GRI (2010) the main question of SR is to what extent do organization impact economically, environmentally and socially on conditions, developments, and trends at the local, regional or global level. In other words, how do they contributes, or aims to contribute in the future, to the improvement or deterioration of their localities. This means that disclosing only trends in individual performance at the expense of performance in relation to broader concepts of sustainability limits the content of the SR and therefore placed on environmental or social resources at the sector, local, regional, or global level (GRI, 2010).

4.3.4 Completeness

Another principle for defining the content of SR is completeness. According to GRI (2010) SR should disclose sufficient material aspects and their Boundaries, to reflect significant economic, environmental and social impacts, and to help stakeholders in the assessment of the organization's performance in the reporting period. The principle of completeness promotes the inclusion of all dimensions of scope, boundary, and time and also used to refer to practices of collecting reasonable and appropriate information (GRI, 2010).

4.4 Principles for Defining Report Quality (PDRQ)

The quality is one of the important characteristics of SR. The PDRQ are used to ensure the quality of information to be presented in the SR. These enable stakeholders of the organisation to make sound and reasonable assessments of performance and take appropriate actions when the SR is finally produced (GRI, 2010). The process of preparing information in a SR should be consistent with PDRQ to achieving transparency. Further, the GRI presented qualitative aspects of information drawing on accounting principles such as balance, clarity, reliability, comparability, timeliness, and accuracy. The various PDRQs are explained in the forgoing paragraphs;

4.4.1 Balance

Balance is of the PDRQs which is used to mean that the SR should reflect both positive and negative aspects of the organization's performance to enable a reasoned assessment of overall performance by all stakeholders (GRI, 2010). The SR should not be seen to be biased in content in presenting the picture of the organization's performance. According to GRI, it should avoid selections, omissions, or presentation formats that are reasonably likely to unduly or inappropriately influence a decision or judgement by the report reader (GRI, 2010).

4.4.2 Clarity

Another PDRQ is the clarity in the SR. This means that organizations are expected to clear and comprehensible information in the SR as well as making the information accessible to all stakeholders (GRI, 2010). What will be the benefit of presenting information which is difficult to understand?

4.4.3 Accuracy

The sustainability information should be sufficiently accurate and detailed for stakeholders to assess the organization's performance according to PDRQ. The sphere of sustainability information (economic, environmental and social) as well as performance indicators can be expressed in many different ways in terms of either qualitative responses or quantitative measurements. The characteristics that determine accuracy vary according to the nature of the information and the user of the information.

4.4.4 Timeliness

The SR should be prepared on regular bases so that sustainability information is made available in time for decision making by the stakeholders. Information is mostly useful when available for use at the right time and form to the stakeholders. The timing here refers both to the regularity of reporting as well as its proximity to the actual events described in the report (GRI, 2010).

4.4.5 Comparability

Another PDRQ is the principle of comparability. This principle state that the organization should select, compile and consistently prepare the SR. The SR should be presented in a way to facilitate easy analysis of the organization's performance over time, and encourage easy comparisons of organisation of equal size and category (GRI, 2010). This principle will enable stakeholders to compare information on the sustainability report against the companies' past performance, its objectives, and, to the degree possible, against the performance of other organizations (GRI, 2010).

4.4.6 Reliability

Finally, another PDRQ is the principle of reliability. The principle state that the organization should gather, record, compile, analyse and disclose sustainability information and processes used in the preparation of a SR should be trusted. This means that the SR should be prepared in such a way those stakeholders will have confidence to establish the veracity of its contents and the extent to which it has appropriately applied the reporting principles (GRI, 2010).

4.4.7 Reporting guidance

Reporting guidance describes actions that can be adopted by the reporting organization when making decisions on what to report on. It also helps to interpret the use of the GRI Reporting Framework (GRI, 2010). There are two forms of reporting guidance namely guidance for defining report content and guidance for setting report Boundary.

4.5 Guidance for Defining Report Content (GDRC) and Guidance for report boundary setting (GRBS)

The GDRC helps organisations to define the content of SR in determining which entities' (e.g., subsidiaries and joint ventures) performance will be included in the report. On the other hand GRBS give guidance that in preparing SR the organisation should include the entities over which they exercises control or significant influence through its relationships with various entities upstream and downstream (GRI, 2010). The terms control and significant influence have been defined as the power to govern the financial and operating policies of an enterprise so as to obtain benefits from its activities and the power to participate in the financial and operating policy decisions of the entity but not the power to control those policies respectively (GRI, 2010). The reporting approach of an organisation will depend on a combination of its control or influence over the entity, and whether the disclosure relates to operational performance, management performance, or narrative/descriptive information (GRI, 2010).

Different relationships involve differing degrees of access to information and the ability to affect outcomes therefore there should be guidance in setting the report boundary. For example, operational information such as emissions data can be reliably compiled from entities under the control of an organization, but may not be available for a joint venture or a supplier. Determining the significance of an entity when collecting information or considering the extension of a boundary depends on the scale of its sustainability impacts. Entities with significant impacts typically generate the greatest risk or opportunity for an organization and its stakeholders, and therefore are the entities for which the organization is most likely to be perceived as being accountable or responsible.

A SR should include in its boundary all entities that generate significant sustainability impacts (actual and potential) and/ or all entities over which the reporting organization exercises control or significant influence with regard to financial and operating policies and practices. The guidelines require standard contents for sustainability reporting regarding the organization's profile, its governance-structures and processes, and the management of sustainability issues including goals and environmental, social and economic performance indicators. To provide a basis for their reports, many organisations have relied on these guidelines in the preparation of SRs (Roca and Searcy, 2012).

Within this sustainability context, what is particularly significant was the principle of completeness that encompassed the "dimensions of scope, boundary and time" (G3, 2010, p.12). The "Scope" referred to the sufficiency of topics covered with reference to stakeholder and related engagement processes. The "Boundary" is taken to include both upstream and downstream entities that firms exercised control or influence and whose performance was presented in the report. The "time" in this context referred to the need to include activities and events in the SR within the period they occurred.

4.6 Standard Disclosures

There are three different types of disclosures which specify the base content that should appear in a SR according to the GRI guidelines. Companies should disclose three levels information types on their SR with respect to (1) Strategy and Profile disclosure (SPD), (2) Management Approach disclosures (MAD) and (3) Performance Indicators disclosures (PID) (see Figure 4.1 and Table 4.4).

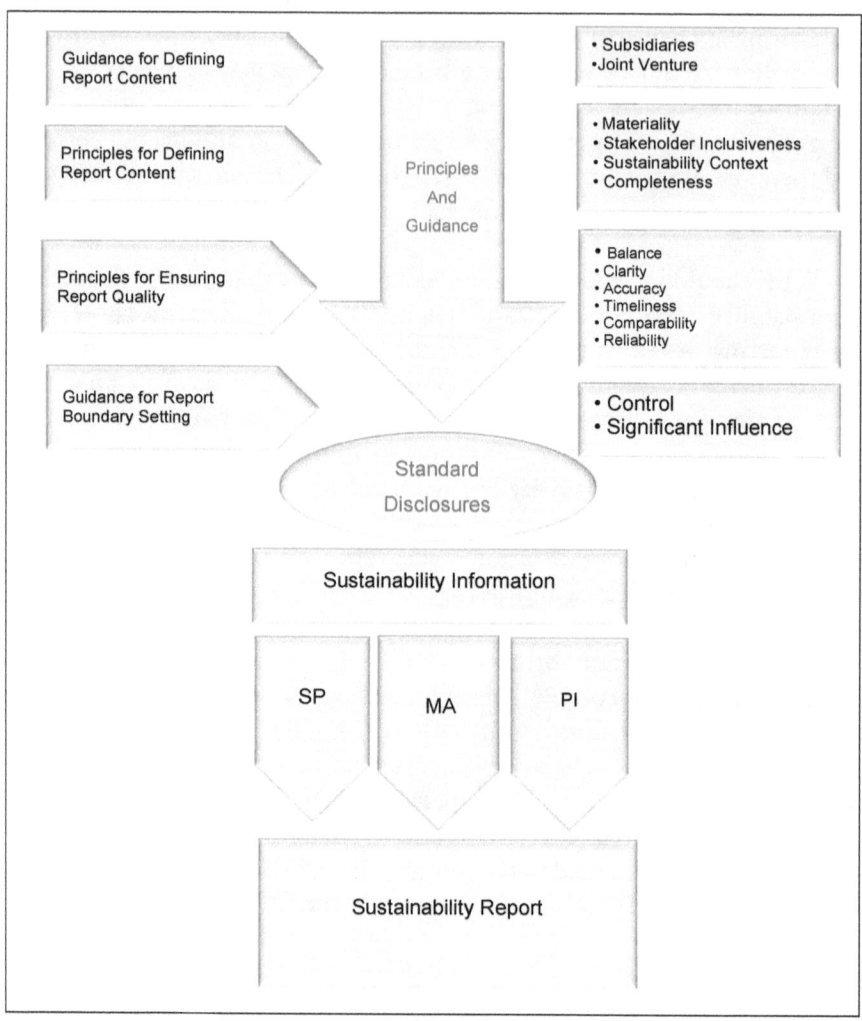

Source: Adapted from GRI guidelines Version 3.0 (2010)
Notes: SP = Strategy and Profile, MA= Management Approach and PI=Performance Indicators
Figure 4.1 GRI Framework principles, guidance and disclosures

Table 4.1 SR indicators in GRI guidelines

Strategy and Profile	Management Approach	Performance Indicators
1. Strategy and Analysis	1. Economic	EC1-EC9
2. Organisation Profile	2. Environmental	EN1-EN30
3. Report Parameters	3. Labour practice and decent Work	LA1-LA15
4.Governance, commitments and Engagement	4. Human Rights	HR1-HR11
	5. Society	SO1-SO10
	6. Product Responsibility	PR1-PR9

Source: Adapted from GRI guidelines Version 3.0 (2006).

4.6.1 Strategy and Profile disclosures

The SPD consist of the main standard disclosure where the company's CEO has to provide an overall "strategy and analysis" that includes an overview of the strategy and risk factors (Joseph, 2012) see Table 4.6. The SPD set the overall context for understanding organizational performance such as its strategy, profile, and governance (GRI, 2010). It provides information on stated organisational aims or values and thus covers corporate recognition of the values of SR (e.g., striving for a reduction in energy consumption). The SPD provide a strategic view on the organisation's sustainability and the general level of the company's CSR activity. Table 4.6 shows the expanded categories of SPD according to GRI guidelines. The regulations require the governance items be disclosed in a separate Section in the annual report. According to the Table areas to be disclosed in terms of this categories includes in broad areas Strategy and Analysis, Organisational Profile, Report Parameters and Governance, Commitments and Engagements. These have been further been broken down in terms of items which can be seen in the Table 4.6.

Table 4.2 Strategy and Profile disclosure

Areas	Items
1. Strategy and Analysis	• Statement from the most senior decision-maker of the organisation (e.g., CEO, chair, or equivalent senior position) about the relevance of sustainability to the organisation and its strategy. • Description of key impacts, risks, and opportunities.
2. Organisational Profile	• Primary brands, products and service. • Operational structure. • Nature of ownership and legal form. • Scale of organisation. • Awards received in the reporting period. • Significant event.
3. Report Parameters	• Reporting period (e.g., fiscal/calendar year) for information provided. • Date of most recent previous report (if any). • Reporting cycle (annual, biennial, etc.) • Contact point for questions regarding the report or its contents. • Boundary of the report (e.g., countries, divisions, subsidiaries, leased facilities, joint ventures, suppliers). • Limitations on the scope or boundary of the report. • Significant changes from previous reporting periods in the scope, boundary, or measurement methods applied in the report.

4.Governance, Commitments and Engagements	• Governance structure of the organisation.
	• Whether the board or management control the governance body.
	• Mechanisms for shareholders and employees to provide recommendations or direction to the highest governance body.
	• Linkage between compensation for members of the highest governance body, senior managers, and executives and the organization's performance.
	• Process to avoid conflict of interest.
	• Process for determining the composition, qualifications, and expertise of top managers including gender and diversity consideration.
	• Internally developed statement of mission, or values, codes of conduct and principles relevant to economic, environmental and social performance.
	• Externally developed economic, environmental, and social charters, principles, or other initiatives to which the organization subscribes or endorses.
	• Address of external precautionary approach.
	• Lists of stakeholders group engaged by organisation.
	• Approach for stakeholder engagement.
	• Key topics and concerns of engagement.

Source: Adapted from GRI guidelines Version 3.0 (2006).

4.6.2 Management approach disclosure

The GRI guidelines provide a structured overview of the base content of MAD which are one of the disclosures in SR also cover how an organization addresses a given set of topics to provide context for understanding performance in a specific area such as economic, environmental, labour practices & decent work, human right, product responsibility and society (GRI, 2010; Bouten et al., 2011). It addresses a given CSR issue by describing the action or practice adopted. The MAD analyse the firm's approach in six areas of social interest and provide a context for understanding the performance in these areas, the detail for which are provided in Table 4.7. The MAD information types are referred to as areas and each of them has items/aspects to be disclosed (Robertson and Nicholson, 1996; Vuontisjärvi, 2006; Bouten et al., 2011) as shown in the Tables 4.7 below. In addition to MAD three categories of indicators, i.e., economic, social, and environmental, the GRI also specifically provides "sector guidance supplements" for use by different industries to address the unique issues in that industry (Joseph, 2012). These sector supplements guides sectors such as Financial Services, Metals, Telecommunications, Public Service and others (Joseph, 2012). It is expected that GRI continues to develop and update sector supplements from time to time to make them relevant to those industries.

Table 4.3 Management approach disclosure

Areas	Items
Economic	• Direct economic impacts • Market presence • Indirect economic impacts
Environment	• Materials • Energy • Water • Biodiversity • Emissions, Effluents and waste • Products and services • Transport
Human Rights	• Investment and procurement practices • Non-discrimination • Freedom of association and collective bargaining • Child labour • Forced and compulsory labour • Security practices • Indigenous rights

Labour practices &	•	Employment
Decent work	•	Labour/management relations
	•	Occupational health and safety
	•	Training and Education
	•	Diversity and equal opportunity
	•	Employee satisfaction
Product	•	Customer health and safety
Responsibility	•	Product and service labelling
	•	Marketing communications
	•	Customer privacy
	•	Customer satisfaction
Society	•	Community
	•	Corruption
	•	Public policy
	•	Anti-competitive behaviour

Source: Adapted from GRI guidelines Version 3.0 (2010).

4.6.2.1 Economic

The economic aspect of the MAD seeks to disclose items such as direct economic impacts, market presence and indirect economic impacts. The economic dimension of sustainability is defined to mean the economic impact of the organization's activities on its stakeholders and on economic systems at local, national, and global levels (GRI, 2010). It does also indicate the economic impacts of the organization throughout society. Financial performance is fundamental to understanding an organization and its own sustainability but this information can normally be found in the financial accounts. The use of organization-specific Indicators in addition to the GRI performance indicators to demonstrate the results of performance against goals is what is often needed by stakeholders of SR. This is considered to be the organization's contribution to the sustainability of a larger economic system.

4.6.2.2 Environmental

According to GRI the environmental dimension of sustainability looks at the organization's impacts on living and non-living natural systems, including ecosystems, land, air, and water (GRI, 2010). Environmental Indicators cover performance related such as material, energy, water, emissions, effluents, waste, biodiversity, environmental compliance, and other relevant information such as environmental expenditure and the impacts of products and services (GRI, 2010), see Table 4.7.

4.6.2.3 Labour Practices & Decent Work

The Labour practices & decent work disclosure according to GRI comprises of employment, labour/management relations, occupational health and safety, training and education, diversity and equal opportunity, and employee satisfaction (GRI, 2010). Some of the internationally recognised standards and instruments that the categories of Labour practices are based are including the following:

- United Nations Universal Declaration of Human Rights;
- United Nations Convention: International Covenant on Civil and Political Rights;
- United Nations Convention: International Covenant on Economic, Social, and Cultural Rights;
- Convention on the Elimination of all Forms of Discrimination against Women (CEDAW);
- ILO Declaration on Fundamental Principles and Rights at Work (in particular the eight core Conventions of the ILO consisting of Conventions 100, 111, 87, 98, 138, 182, 29, 105); and
- The Vienna Declaration and Programme of Action.
- the ILO Tripartite Declaration Concerning Multinational Enterprises
- and Social Policy and the Organisation for Economic Cooperation and Development (OECD) Guidelines for Multinational Enterprises (GRI, 2010).

4.6.2.4 Human Rights

There is an increasing consensus globally, that mandate organizations to have the responsibility to respect human rights and to report on the extent to which processes have been implemented, on incidents of human rights violations and on changes in the stakeholders' ability to enjoy and exercise their human rights, occurring during the reporting period (GRI,

2010). From the Table 4.7 the human rights disclosures included are non-discrimination, gender equality, and freedom of association, collective bargaining, child labour, forced and compulsory labour, and indigenous rights. Furthermore there is international legal framework for human rights which comprises of a body of law made up of treaties, conventions, declarations and other instruments serve as a point of reference for any organisation in the preparation of SR. According to GRI (2010), the corner stone of human rights is the International Bill of Rights which is formed by three instruments as follows:

1) the Universal Declaration of Human Rights (1948);
2) the International Covenant on Civil and Political Rights (1966); and
3) the International Covenant on Economic, Social and Cultural Rights (1966).

4.6.2.5 Society

With reference to Society performance indicators organisations are expected to disclose the impacts of their activities on the local communities in which they operate, in relation to the risks that may arise from their interactions with other social institutions and how they are managed and mediated. According to GRI the society information such as risks associated with bribery and corruption, undue influence in public policy-making, and monopoly practices are to be reported in the SR (GRI, 2010). There is ongoing debate that collective community rights, indigenous and tribal peoples have collective rights which are recognized by ILO Conventions 107 and 169 and the UN Declaration on Indigenous Rights (GRI, 2010). Community members have individual rights based on the following:

• Universal Declaration of Human Rights;
• International Covenant on Civil and Political Rights;
• International Covenant on Economic, Social and Cultural Rights; and
• Declaration on the Right to Development.

The community have a right to free, prior and informed consultation in order to seek consent is a fundamental right expressly recognized in the reference points above.

4.6.2.6 Product Responsibility

Product Responsibility was defined by GRI to include the aspects of a SR of organization's products and services that directly affect customers, namely, health and safety, information and labelling, marketing, and privacy (GRI, 2010). SR of an organisation is expected to disclose the internal procedures and the extent to which these procedures are not complied regarding performance relevant to the Product Responsibility aspects by the organisation.

4.6.3 Performance Indicators

The Performance Indicators (PIs) are developed to provide more detail in each of the six management approach disclosure (MAD) areas while the sector supplements provide the measures, along with indicators that are specific to the industry (Joseph, 2012; GRI, 2010). The sustainability PIs are organized by economic, environmental, and social categories. Social Indicators are further categorized by Labour, Human Rights, Society, and Product Responsibility. PIs elicit comparable information on MAD performance of the organization and CSR achievements by providing quantitative measures of CSR performance (GRI. 2010). There are two categories of PI that should be disclosed namely set of Core and Additional Performance Indicators.

Core Indicators are intended to identify generally applicable Indicators and are assumed to be material for most organizations which they are expected to report on unless they are deemed immaterial based on GRI Reporting Principles. According to the G3 guidelines, core indicators are developed through the multi-stakeholder process and are assumed to apply to all firms (Joseph, 2012). Additional Indicators also address the emerging topics that may be material for some organizations, but are not material for others. For example where Sector Supplements exist, the Indicators should be treated as Core Indicators (GRI, 2010). Reporting organizations are encouraged to follow this structure in preparing their SRs, however, other formats may be chosen.

The GRI G3 guidelines released in 2006 outline several reporting parameters, reporting principles, and standard disclosures, including a list of 79 PIs. The PIs are made-up of 9 economic indicators, 30 environmental indicators, and about 40 social indicators that have been further categorized into labour practices & decent work, human rights, society, and product responsibility. For example an area 'Economic' which has one of its items/ aspect as 'Direct economic impacts' may have EC1 which stands for 'Direct economic value generated and distributed, including revenues, operating costs, employee compensation, donations and other community investments, retained earnings, and payments to capital providers and governments' as one of the PIs (GRI guidelines Version 3.0, 2010), see Tables 4.8, 4.9, 4.10, 4.11, 4.12 and 4.13 below. GRI also has categories of reporting to capture different stages in SR preparation by organisations, where the B category is for those that report at least 20 measures and the A category for those who addressed all measures (either reporting the indicators or indicating that they are not applicable) (GRI guidelines Version 3.0, 2010; Joseph, 2012). This means that, there is considerable flexibility for organisations in complying with the measurement criteria using GRI guidelines.

The core G3 guidelines which were updated in 2011 (version 3.1, GRI, 2011b) are supported by numerous sector supplements and country-specific annexes. The GRI as the most widely known set of voluntary guidelines for SR have it aim of disclosing of environmental, social and governance performance" (GRI, 2011a). The main underlying assumptions of GRI are the strengthening of civil private regulation, CSR, and collaborative governance (Brown et al., 2009a).

Table 4.4 Economic Performance Indicators

Economic Performance Indicators		
Aspect:		**Core**
Economic Performance	**EC1**	Direct economic value generated and distributed, including revenues, operating costs, employee compensation, donations and other community investments, retained earnings, and payments to capital providers and governments.
	EC2	Financial implications and other risks and opportunities for the organization's activities due to climate change.
	EC3	Coverage of the organization's defined benefit plan obligations.

	EC4	Significant financial assistance received from government.
Market Presence	EC5	Range of ratios of standard entry level wage by gender compared to local minimum wage at significant locations of operation.
	EC6	Policy, practices, and proportion of spending on locally-based suppliers at significant locations of operation.
	EC7	Procedures for local hiring and proportion of senior management hired from the local community at significant locations of operation.
Indirect Economic Impacts	EC8	Development and impact of infrastructure investments and services provided primarily for public benefit through commercial, in-kind, or pro bono engagement.
	EC9	Understanding and describing significant indirect economic impacts, including the extent of impacts.

Source: (version 3.1, GRI, 2011b)

Table 4.5 Environment Performance Indicators

Environment Performance Indicators		
Aspect:		**Core**
Materials	EN1	Materials used by weight or volume.
	EN2	Percentage of materials used that are recycled input materials.
Energy	EN3	Direct energy consumption by primary energy source.
	EN4	Indirect energy consumption by primary source.
	EN5	Energy saved due to conservation and efficiency improvements.

	EN6	Initiatives to provide energy-efficient or renewable energy-based products and services, and reductions in energy requirements as a result of these initiatives.
	EN7	Initiatives to reduce indirect energy consumption and reductions achieved.
Water	**EN8**	Total water withdrawal by source.
	EN9	Water sources significantly affected by withdrawal of water.
	EN10	Percentage and total volume of water recycled and reused.
Biodiversity	**EN11**	Location and size of land owned, leased, managed in, or adjacent to, protected areas and areas of high biodiversity value outside protected areas.
	EN12	Description of significant impacts of activities, products, and services on biodiversity in protected areas and areas of high biodiversity value outside protected areas.
	EN13	Habitats protected or restored.
	EN14	Strategies, current actions, and future plans for managing impacts on biodiversity.
	EN15	Number of IUCN Red List species and national conservation list species with habitats in areas affected by operations, by level of extinction risk.
Emissions, Effluents, and Waste	**EN16**	Total direct and indirect greenhouse gas emissions by weight.
	EN17	Other relevant indirect greenhouse gas emissions by weight.
	EN18	Initiatives to reduce greenhouse gas emissions and reductions achieved.
	EN19	Emissions of ozone-depleting substances by weight.
	EN20	NOx, SOx, and other significant air emissions by type and weight.

	EN21	Total water discharge by quality and destination.
	EN22	Total weight of waste by type and disposal method.
	EN23	Total number and volume of significant spills.
	EN24	Weight of transported, imported, exported, or treated waste deemed hazardous under the terms of the Basel Convention Annex I, II, III, and VIII, and percentage of transported waste shipped internationally.
	EN25	Identity, size, protected status, and biodiversity value of water bodies and related habitats significantly affected by the reporting organization's discharges of water and runoff.
Products and Services	**EN26**	Initiatives to mitigate environmental impacts of products and services, and extent of impact mitigation.
	EN27	Percentage of products sold and their packaging materials that are reclaimed by category.
Compliance	**EN28**	Monetary value of significant fines and total number of non-monetary sanctions for noncompliance with environmental laws and regulations.
Transport	**EN29**	Significant environmental impacts of transporting products and other goods and materials used for the organization's operations, and transporting members of the workforce.
Overall	**EN30**	Total environmental protection expenditures and investments by type.

Source: (version 3.1, GRI, 2011b)

The GRI aims to achieve uniform and consistent reporting on sustainability performance, allowing this to be as routine and comparable as financial reporting (Isaksson and Steimle, 2009). But GRI like any initiative have some strengths and weaknesses. One of such strengths has been the fact that it has a multi-stakeholder features in the process of SR preparation (Brown et al., 2009a). Furthermore, Brown et al. (2009a) argue that the GRI have many characteristics of an established institution, including widespread uptake, legitimacy, emergence of new business activities, and emergence of competitive pressures related to the GRI, among others. The GRI has also been criticized on many occasions such as, confusion over its scope, the lack of a requirement for independent verification of the report, and the fact that different levels of application permit selective reporting on the performance indicators (Moneva et al., 2006; Roca and Searcy, 2012).

Table 4.6 Labour Practices & Decent Work Performance Indicators

Labour Practices & Decent Work Performance Indicators		
Aspect:		**Core**
Employment	**LA1**	Total workforce by employment type, employment contract, and region, broken down by gender.
	LA2	Total number and rate of new employee hires and employee turnover by age group, gender, and region.
	LA3	Benefits provided to full-time employees that are not provided to temporary or part-time employees, by significant locations of operation.
	LA15	Return to work and retention rates after parental leave, by gender.
Labour/ Management Relations	**LA4**	Percentage of employees covered by collective bargaining agreements.
	LA5	Minimum notice period(s) regarding significant operational changes, including whether it is specified in collective agreements.
Occupational Health and Safety	**LA6**	Percentage of total workforce represented in formal joint management-worker health and safety committees that help monitor and advice on occupational health and safety programs.

	LA7	Rates of injury, occupational diseases, lost days, and absenteeism, and total number of work-related fatalities, by region and by gender.
	LA8	Education, training, counselling, prevention, and risk-control programs in place to assist workforce members, their families, or community members regarding serious diseases.
	LA9	Health and safety topics covered in formal agreements with trade unions. Health and safety topics covered in formal agreements with trade unions.
Training and Education	**LA10**	Average hours of training per year per employee, by gender, and by employee category.
	LA11	Programs for skills management and lifelong learning that support the continued employability of employees and assist them in managing career endings.
	LA12	Percentage of employees receiving regular performance and career development reviews, by gender.
Diversity and Equal Opportunity	**LA13**	Composition of governance bodies and breakdown of employees per employee category according to gender, age group, minority group membership, and other indicators of diversity.
Equal Remuneration for Women and Men	**LA14**	Ratio of basic salary and remuneration of women to men by employee category, by significant locations of operation.

Source: (version 3.1, GRI, 2011b)

The main question that may be asked by an interested reader of a SR is how sustainable is the MCGs and whether the company is improving within certain period of time. This means that the company has to identify its main sustainability aspects and create relevant indicators for monitoring the position and development. Many studies have provided an indication of the wide variety of approaches to SR preparation (Roca and Searcy, 2012) that will provide the sustainable information required. These studies have provided needed insight into what goes into the preparation of SR with the direct consideration of the content, scope, and structure of the reports. In all these there has been one important component of SRs that have been widely overlooked (Roca and Searcy, 2012).

Table 4.7 Human Rights Performance Indicators

Human Rights Performance Indicators		
Aspect	Core	
Investment and Procurement Practices	HR1	Percentage and total number of significant investment agreements and contracts that include clauses incorporating human rights concerns, or that have undergone human rights screening.
	HR2	Percentage of significant suppliers, contractors, and other business partners that have undergone human rights screening, and actions taken.
	HR3	Total hours of employee training on policies and procedures concerning aspects of human rights that are relevant to operations, including the percentage of employees trained.
Non-discrimination	HR4	Total number of incidents of discrimination and corrective actions taken.
Freedom of Association and Collective Bargaining Child Labour	HR5	Operations and significant suppliers identified in which the right to exercise freedom of association and collective bargaining may be violated or at significant risk, and actions taken to support these rights.

	HR6	Operations and significant suppliers identified as having significant risk for incidents of child labour, and measures taken to contribute to the effective abolition of child labour.
Forced and Compulsory Labour	HR7	Operations and significant suppliers identified as having significant risk for incidents of forced or compulsory labour, and measures to contribute to the elimination of all forms of forced or compulsory labour.
Security Practices	HR8	Percentage of security personnel trained in the organization's policies or procedures concerning aspects of human rights that are relevant to operations.
Indigenous Rights	HR9	Total number of incidents of violations involving rights of indigenous people and actions taken.
Assessment	HR10	Percentage and total number of operations that have been subject to human rights reviews and/or impact assessments.
Remediation	HR11	Number of grievances related to human rights filed, addressed, and resolved through formal grievance mechanisms.

Source: (version 3.1, GRI, 2011b)

According to Adams and Frost (2008) there is the need to include key PIs in SRs but this, they say have not been adequately explored. In view of this fact, there have been a many studies that have looked at the structure and content of SRs. This has been one of the justifications for this study. For example, several studies have been made of the content of Canadian SRs (CBSR, 2008; Davis and Searcy, 2010; Stratos, 2008). KPMG and SustainAbility have also analysed the content of SRs around the world on several occasions. In the most recent work on the content of SRs by KPMG was completed in 2008 (Slater, 2008) while that of SustainAbility's completed in 2006 (Beloe et al., 2006). Many researches have been conducted over the last decade at the National-level explicitly noted that the most commonly disclosed are social indicators issues such as "workplace health and safety policies and measures, employee education and skill management, and the benefits that employees receive from the organisation (Langer, 2006; Sobhani et al., 2009; (Skouloudis et al., 2010; Vormedal and Ruud, 2009; Hedberg and von Malmborg, 2003; Stiller and Daub, 2007; Ratanajongkol et al., 2006; Roca and Searcy, 2012).

Table 4.8 Society Performance Indicators

Society Performance Indicators		
Aspect		**Core**
Local Communities	**SO1**	Percentage of operations with implemented local community engagement, impact assessments, and development programs.
	S09	Operations with significant potential or actual negative impacts on local communities.
	SO10	Prevention and mitigation measures implemented in operations with significant potential or actual negative impacts on local communities.
Corruption	**SO2**	Percentage and total number of business units analysed for risks related to corruption.
	SO3	Percentage of employees trained in organization's anti-corruption policies and procedures.
	SO4	Actions taken in response to incidents of corruption.
Public Policy	**SO5**	Public policy positions and participation in public policy development and lobbying.
	SO6	Total value of financial and in-kind contributions to political parties, politicians, and related institutions by country.
Anti-Competitive Behaviour	**SO7**	Total number of legal actions for anticompetitive behaviour, anti-trust, and monopoly practices &their outcomes.

Compliance	SO8	Monetary value of significant fines and total number of non-monetary sanctions for noncompliance with laws and regulations.

Source: (version 3.1, GRI, 2011b)

Table 4.9 Product Responsibility Performance Indicators

Product Responsibility Performance Indicators

Aspect:		Core
Customer Health and Safety	**PR1**	Life cycle stages in which health and safety impacts of products and services are assessed for improvement, and percentage of significant products and services categories subject to such procedures.
	PR2	Total number of incidents of non-compliance with regulations and voluntary codes concerning health and safety impacts of products and services, by type of outcomes.
Product and Service Labelling	**PR3**	Type of product and service information required by procedures and percentage of significant products and services subject to such information requirements.
	PR4	Total number of incidents of non-compliance with regulations and voluntary codes concerning product and service information and labelling, by type of outcomes.
	PR5	Practices related to customer satisfaction, including results of surveys measuring customer satisfaction.
Marketing Communications	**PR6**	Programs for adherence to laws, standards, and voluntary codes related to marketing communications, including advertising, promotion, and sponsorship.

	PR7	Total number of incidents of non-compliance with regulations and voluntary codes concerning marketing communications, including advertising, promotion, and sponsorship, by type of outcomes.
Customer Privacy	PR8	Total number of complaints regarding breaches of customer privacy and losses of customer data.
Compliance	PR9	Monetary value of significant fines for non-compliance with laws and regulations concerning the provision and use of products and services.

Source: (version 3.1, GRI, 2011b)

Gallego (2006) conducted an analysis of the use of the indicators suggested by the 2002 version of the GRI in 19 Spanish companies. The study found that the most frequently reported environmental indicators were "related to energy, water, biodiversity and emissions, effluents and waste" while the most frequently reported social indicators were "related to labour practices & decent work, strategy and management, non-discrimination, freedom of association, child labour and forced and compulsory labour" (Gallego, 2006). All companies in the study were found to have reported on the indicator "net sales". In another studies by Skouloudis and Evangelinos (2009) on analysis of economic, environmental, and social performance disclosures, they concluded that the most frequently disclosed economic indicators were "net sales; costs of all purchased goods, materials, and services; total payroll; and benefits". They also indicated that all of the reports contained a "discussion of the donations and charitable contributions that the organizations made during the reporting period" (Skouloudis and Evangelinos, 2009). With respect to the most commonly disclosed environmental indicators according to Skouloudis and Evangelinos (2009) were energy and water consumption, carbon dioxide emissions, and internal initiatives to improve energy efficiency.

Another important factor that needs to be considered is ability of the companies to determine the relevant performance indicators to be disclosed for the benefit of their stakeholders. After identifying stakeholders and determining their concerns, companies must determine the relevant indicators that would correspond to the stakeholders and their needs. This goal was accomplished with the help of GRI guidelines which provides guidance and principles for defining content, quality of reports and performance indicators or measures to be included in the SR. According to GRI guidelines, it is very significant in defining content that integrated sustainability concepts to include stakeholder inclusiveness and sustainability context. The term stakeholder inclusiveness reflects the inclusion of entities that have a legitimate claim over the organization under law or international conventions.

In the preparation of the SR it is expected that the MCGs must indicate how they have responded to the "reasonable expectations and interests" of such stakeholders, including those that are "unable to articulate their views on a report and are represented by proxies." With reference to GRI guidelines initiatives, the sustainability context should extends the SR to include how "an organization contributes, or aims to contribute in the future, to the improvement or deterioration of economic, environmental, or social conditions, developments, and trends at the local, regional, or global level (G3, p. 11).

The discussions of the stakeholder inclusiveness also led to the question of whether the principles proposed by the GRI made provisions for the goals of stakeholders. Despite the calls for transparency in the GRI principles, it does not seem to address the ambiguities underlying sustainability. For example, the "reasonable expectations of stakeholders" (G3; p. 7) was mentioned in the guidelines but no explanation was given of such expectations (Mitchell et al., 1997). As noted earlier, the guidelines also highlighted the need to make provisions for "stakeholders who are unable to express their views on a report and whose concerns are represented by proxies" (G3; p. 10). According to Mitchell et al. (1997), such stakeholders are in line with what they considered as "dependent" stakeholders, who have legitimate stakes in the activities of the company, but lack the power to exercise their stakes (Mitchell et al., 1997). The lack of norms as well as the criteria for identifying stakes of these stakeholders affects the extent to which companies will disclose such stakes to address the stakeholders' expectations.

The analysis above leads to the question what kind of performance indicators does the companies and for that matter whether the MCGs prepare their SR according to GRI guidelines in order to achieve the goal of sustainability reporting. The discretionary nature of the application of principles of the GRI guidelines extends to measurement contained in the "standard disclosure" that sought to disclose "management approach" to risks and strategies. It is expected that companies need to be more focused on social as opposed to economic or even environmental to where they operate. This means they have to adopt measures that meet these specific goals in order to have greater in impact. Furthermore, measures that might be chosen by companies and use of different types of indicators will differ based on the country, the type of company activities and social/legal structure.

In spite of all the efforts made to understand the roots of sustainability, the principles are not integrated in a cohesive manner to address the challenges of sustainability. GRI extends the traditional accounting lens into the stakeholder theory, with more latitude and the development of measures to provide companies opportunity to be aware of their role in locality without seriously examining areas of ambiguities or the necessity for sustainability. The term sustainable development is more open to many interpretations without addressing the relationship of such a goal to companies (Moneva et al., 2006). According to Moneva et al. (2006), the simplistic perception of sustainable development made GRI to obscure the long-term perspective of sustainability of construction of a set of performance indicators rather than considering all aspects of sustainability in its entirety. The question that can be asked on the application of the stakeholder theory and its ability to address stakeholder concerns is that, in the absence of regulation, what sort of performance indicators is being disclosed by the MCGs?

4.7 Summary

The GRI Reporting Framework is intended to provide a generally accepted framework for reporting on an organization's economic, environmental, and social performance. The chapter was used to present principles of reporting and defining the content of sustainability reports. Guidance for defining the report content and the standard disclosure was also included in this chapter where strategy and profile disclosure, management approach disclosures and performance indicators were presented.

CHAPTER 5

GRI Indicators Disclosures by Mining companies in Ghana

5.1 Introduction

This chapters analysed the data gathered on the kind of performance indicators reported by mining companies in Ghana. The Sections 5.2 to 5.7 were used for the analyses of Economic indicators, Environmental indicators, Human Rights indicators, Labour practice and decent work indicators, Product responsibility indicators and Society indicators. The trend and the company level disclosures were presented in Section 5.8. The findings on the research questions are summed up in the conclusion section of this chapter where the main research question was answered.

5.2 GRI Indicators in Sustainability Reports

Sustainability issues are increasingly gaining importance among corporations and their stakeholders in the world over (Roca and Searcy, 2012). Over the last three decades, the many corporations have been preparing SR and made it available to the public, however, despite the proliferation of these reports, questions still remain on the kind of information the SR should contain as well as its structured (Davis and Searcy, 2010; Roca and Searcy, 2012). The preparation of SR in most jurisdictions remains voluntary and Ghana is of no exception. The voluntary nature also makes it difficult to have a uniform format for the presentation of the reports. To alleviate such difficulties, numerous reporting guidelines have been published to guide corporations in the development of sustainability reports, most notably by the Global Reporting Initiative (GRI, 2006; Roca and Searcy, 2012). GRI Guidelines focuses on how to report and prepare a sustainability report (Guthrie and Farneti, 2008). The Guidelines have been created in order to establish a framework to apply to companies, non-profit organisations, public agencies and others (GRI, 2006a, p.2), who want to provide voluntary sustainability reports.

As has been indicated above, GRI defines sustainability reporting largely through a triple bottom line (TBL) perspective (GRI, 2002, p.9) namely social, environmental and economic information. The term is also used synonymously with citizenship reporting, social reporting, triple bottom line reporting and other terms that encompass the economic, environmental and social aspects of an organisation's performance (GRI, 2005, p.16). The GRI Guidelines have been taken up by organisations worldwide and as at March 2005, by over 630 organisations in 51 different countries (GRI, 2005, p. 4), while at November 2006, nearly 1,000 organisations in over 60 countries have stated that they are using the GRI Reporting Framework (GRI, 2006b; Guthrie and Farneti, 2008).

According to the GRI guidelines and many other initiatives, SR should be prepared to contain a description of the organization, its sustainability vision, its objectives towards sustainability, and a series of indicators illustrating the performance of the organization, among other issues (GRI, 2006; Roca and Searcy, 2012). There is a growing body of research on what motivate organisations to develop the structure and content of SR (Roca and Searcy, 2012). According to Adams and Frost (2008) as quoted by Roca and Searcy, (2012), "a considerable doubt has been cast on the extent to which many sustainability reports accurately and completely portray corporate social and environmental impacts." Generally, there are many potential reasons associated with this inaccuracy of reportage of the SR. One of the main reasons that have been attributed may be the emphasis placed on qualitative information in most SR of many organisations (Roca and Searcy, 2012). In Ghana SR of many organisations are perceived to contain a lot of pictorial elements which are mainly qualitative information. According to Roca and Searcy, 2012 few studies have been made on the performance indicators (PI) used to convey quantitative information in SR. These PIs are said to represent the concrete data on the organisation's performance in relation to sustainability (Daub, 2007). Therefore the PIs are considered to be as important as the qualitative part of SR.

Studies on PIs especially its development process has been made by several authors (Roca and Searcy, 2012; Daub, 2007; Chee Tahir and Darton, 2010). Most works done on PIs are centred around the "triple bottom line" namely economic, environmental, and social performance (Elkington, 1998) and examples of sustainability PIs are widely available in most studies (see, for example, Veleva and Ellenbecker, 2001; Azapagic, 2004; Nordheim and Barrasso, 2007; Searcy et al., 2007). The GRI guidelines has 79 indicators and are organized around this triple bottom line, with the social dimension sub-divided into labour practices, human rights, society, and product responsibility indicators (GRI, 2006; Roca and Searcy, 2012). In spite of these guidelines the extent to which the suggested PIs are actually used in SR is generally questionable according to Roca and Searcy, (2012) and MCGs are. The purpose of this Chapter of the study is to explore the PIs disclosed in SR of the MCGs. The research question of this part of the study intended to answer is "What indicators are currently being disclosed in SRs of the MCGs?" The results from the study can provide insights that may form a basis for further research in other areas, although caution must be exercised in applying them due to the fact that it pertain to only one sector in Ghana. The results are presented below in several subsections below to examine the use of the GRI indicators in MCGs' sustainability reports.

The object of the research project is to draft a country report on mining about the sort of performance indicators they are disclosing. As indicated in the introductory Chapter, this study is restricted to the top 10 mining companies in Ghana. The main reason being that experience to-date shows that a much larger share of the responsibility for global problems such as the pollution of the environment or social inequality is placed on the shoulders of large companies compared to small-to-medium-sized companies which are normally under intense pressure by their stakeholders to behave well in their day to day activities (Daub, 2007). Most of these small-to-medium-sized companies are locals and are expected that very few sustainability reports would be published by them. The results of the many international studies suggest that sustainability reports are mainly published by multinational companies during a primary phase (Santos, 2001; 2003; SustainAbility, 2002; Daub, 2007). This gives justification for the use of annual reports of the large commercial mining companies in the study.

The study, as noted earlier, used GRI guidelines (GRI, 2002) as the method of evaluation as regards to many previous systems (Santos, 2001; 2003; SustainAbility, 2002). According to Alazzani and Wan-Hussin (2013) recent study on GRI indicators is focused on non-financial corporate reporting in specific industries, such as the petrochemical industry (Samuel et al., 2013), the forest industry (Toppinen et al., 2012), mining (Kraut et al., 2012), and the five most polluting industries: pulp and paper, chemicals, oil and gas, metals and mining, and utilities (Clarkson et al., 2008; Alazzani and Wan-Hussin, 2013). In assessing the extent to which MCGs followed the GRI Sustainability Reporting Guidelines in the study, GRI indicators was chosen because they provide the most comprehensive reporting framework of principles, standard disclosures, indicators and reporting protocols, augmented by a series of sector-specific supplements and resource documents (Alazzani and Wan-Hussin, 2013).

Another argument in favour of the choice of a GRI-based criteria catalogue is based on the similar studies of Moore (2002) and Moore and Robson (2002), where they also adopted GRI guidelines (see Figure 4.1 for GRI Framework, principles, guidance and disclosure) in the analysis of corporate social and financial performance of the UK supermarket industry. Furthermore, GRI framework is globally recognized as the most widely used sustainability reporting tool; the performance indicators listed therein are used to measure and report an organization's economic, environmental, and social performance, in what is also known as 'triple bottom line' or 'People, Planet, Profit' reporting (GRI, 2012; Alazzani and Wan-Hussin, 2013). The first set of formal Sustainability Reporting Guidelines was published in June 2000 to cover three dimensions: environment, economic, and social.

In the view of Morhardt et al. (2002, p.9; 215), the GRI Sustainability Reporting Guidelines 2000 "are considered the most detailed, comprehensive, and prescriptive guidelines to-date" and sticking to it as a guide in the preparation of SR would lead to a tremendous performance by any company. The updated version in 2006 further added advantage to this challenge. In spite of the extensive nature as well as supporting the idea of report standardization of the content SR, GRI guidelines do not mandate the organizations to fully fulfil or handle all topics. Due to the fact that the GRI guidelines are made to fit all types of companies, companies are free to use the guidelines in any way they choose Daub, (2007). According to Daub (2007) this freedom of application can both be seen as strengths as well as the weaknesses of the GRI guidelines. One of such weakness is that, the topics and indicators are written in a fairly general way, making the implementation for many companies difficult (Daub, 2007). As agreed by many studies, it is the most commonly used reporting guidelines by international companies (Daub, 2007).

Sustainability reports were considered as a single construct, in which companies perform environmental activities and disclose such activities. Therefore, the presence (1) or absence (0) of certain words and concepts in texts covering corporate disclosures within SR were detected using GRI guidelines because of its provision of clear definitions for each indicator, which make it easy and accurate to assess the companies' environmental performance (Alazzani and Wan-Hussin, 2013). All the MCGs reported using the GRI G3 guidelines in 2008 and 2012 in the preparation of their SRs. All the 98 performance indicators including listed in the GRI G3 guidelines were used at least once. Tables 5.1 to 5.6 provide the GRI performance indicators reported in the various SRs of the MCGs within the 2008 and 2012 and further details on the use of the GRI indicators are provided below. The detailed item-wise disclosure of corporate sustainability performance indicators of the mining companies in Ghana has been presented under the management approach themes: (i) economic performance indicator disclosure, (ii) environmental performance indicator disclosure, (iv) human right performance indicators, (v) labour practices & decent work performance indicators, product responsibility performance indicators and (vi) society performance indicator disclosure.

Table 5.1 MCGs and their GSE Listings

MCGs	Type of Mineral	Listing status on GSE
AngloGold Ashanti	Gold	Listed
Chirano	Gold	Not listed
Golden Star Resources	Gold	Listed
Newmont Ghana Limited	Gold	Not listed
Tarkwa Goldfields Limited	Gold	Not listed
Ghana Manganese Company	Manganese	Not listed
Ghana Bauxite	Bauxite	Not listed

Endeavour Mining Resources	Gold	Not listed
Noble Mineral Resources	Gold	Not listed
Persues	Gold	Not listed

Source: Author

5.3 Economic Indicators (EC)

Corporate economic sustainability is used to measure the economic outcomes of an organization's activities and the impact on their stakeholders (GRI, 2006; Sobhani et al., 2012).The economic performance of an organization is fundamental to understanding the organization and its basis for sustainability due to the fact that an organization may be financially viable, but may have been achieved by creating significant externalities that impact other stakeholders (Sobhani et al., 2012). Table 5.2 shows that the economic performance indicators disclosure indices of 80 percent of the research instrument items have been disclosed in the SR by the MCGs. This means that majority of indicators have been disclosed by the MCGs. With respect to the item-wise disclosure in the SR, information concerning economic performance indicators disclosures that are expected to be included in the SR are Economic Performance, Market Presence, Indirect Economic and Impacts of the activities by the organisations. Table 5.2 presents the frequency of use of the GRI economic indicators by the MCGs in the period under consideration. The GRI's economic indicators were widely reported in SR of many MCGs during that period. For example, it can be seen from Table 5.2 that all the MCGs reported on EC1 ("Direct economic value generated and distributed, including revenues, operating costs, employee compensation, donations and other community investments, retained earnings, and payments to capital providers and governments"), EC4, ("significant financial assistance received from government") and EC7 ("Procedures for local hiring and proportion of senior management hired from the local community at significant locations of operation").

Table 5.2 Disclosures of the Global Reporting Initiative (GRI) G3 Economic performance indicators

	AGA		CHI		GSR		NGL		TGF		GMC		GBL		EML		NMR		AMR		Total	
	08	12	08	12	08	12	08	12	08	12	08	12	08	12	08	12	08	12	08	12	08	12
EC1	1	1	1	1	0	1	1	1	1	1	1	1	1	1	1	1	1	1	1	1	9	10
EC2	0	1	1	1	0	1	0	1	0	1	0	1	1	1	0	1	0	1	1	0	3	9
EC3	0	1	1	1	1	1	0	1	0	1	0	1	0	1	0	1	0	1	0	0	2	9
EC4	0	1	0	1	0	1	0	1	1	1	0	1	0	1	0	1	0	1	0	1	1	10
EC5	1	1	0	1	1	1	1	1	1	1	0	1	0	1	0	1	0	1	0	0	4	9
EC6	0	1	0	1	0	1	0	1	0	1	0	1	1	1	0	1	0	1	0	0	1	9
EC7	1	0	1	0	0	0	1	1	0	1	0	1	1	0	1	0	1	1	1	1	7	5
EC8	1	1	0	1	0	0	1	1	1	1	1	0	0	1	0	0	0	1	0	1	4	7
EC9	1	1	1	1	1	1	1	1	0	1	0	0	0	1	0	1	0	0	0	1	4	8
MM1	0	0	0	1	0	0	0	1	0	1	0	1	0	1	0	0	0	0	0	0	0	5
MM2	0	1	0	1	0	1	0	1	0	1	0	1	0	1	0	0	0	0	0	0	0	7
ECPDI	5	9	5	10	3	8	5	11	4	11	2	9	4	10	2	7	2	8	3	5	35	88

Source: Author

Notes: AGA = AngloGold Ashanti, CHI = Chirano, GSR = Golden Star Resources, NGL=Newmont, TGF =Goldfields, GMC = Ghana Manganese Company, GBL = Bauxite, EML = Endeavour Mineral Resources, NMR = Noble Minerals Resources, AMR = Persues Minerals Resources.

This is followed by EC2 ("Financial implications and other risks and opportunities for the organization's activities due to climate change") EC3 ("Coverage of the organization's defined benefit plan obligations"), EC5 ("Range of ratios of standard entry level wage by gender compared to local minimum wage at significant locations of operation") and EC6 ("Policy, practices, and proportion of spending on locally-based suppliers at significant locations of operation") where 9 out of 10 MCGs reported on them in the SRs. With regards to EC8 ("Development and impact of infrastructure investments and services provided primarily for public benefit through commercial, in-kind, or pro bono engagement") only 7 MCGs made mention of it in their report of GRI economic indicators. Only 3 MCGs reported on EC9 ("understanding and describing significant indirect economic impacts, including the extent of impacts") according to the results in Table 5.2.

5.4 Environmental Performance Indicators (EN)

Environmental performance indicators concern an organization's impact on living and non-living natural systems, including ecosystems, land, air, and water as well as covering performance related to input (e.g., material, energy, water) and output (e.g., emissions, effluents, waste) (Sobhani et al., 2012). The environmental performance indicators are said to be very important as far as mining is concern due the devastating consequences of mining operations. According to GRI, EN information aspects to be reported on broadly include Materials, Energy, Water, Biodiversity, Emissions, Effluents, and Waste, Products and Services, Compliance, Transport and Overall (GRI, 2006). The frequency of use of the GRI environmental indicators by MCGs is can be seen in Table 5.3. The GRI environmental indicators that are fully disclosed in the sustainability reports are EN1 ("materials used by weight or volume"), EN2 ("percentage of materials used that are recycled input materials"), EN3 ("Direct energy consumption by primary energy source"), EN16 ("Total direct and indirect greenhouse gas emissions by weight") and EN17 ("other relevant indirect greenhouse gas emissions by weight"). The least frequently disclosed EN indicators (reported by 2 MCGs) were EN24 ("Weight of transported, imported, exported, or treated waste deemed hazardous under the terms of the Basel Convention Annex I, II, III, and VIII, and percentage of transported waste shipped internationally"). Two indicators were reported by 9 MCGs namely EN11 ("location and size of land owned, leased, managed in, or adjacent to, protected areas and areas of high biodiversity value outside protected areas") and EN12 ("Description of significant impacts of activities, products, and services on biodiversity in protected areas and areas of high biodiversity value outside protected areas"). EN indicators such as EN19 ("emissions of ozone-depleting substances by weight"), EN20 ("NOx, SOx, and other significant air emissions by type and weight"), EN21 ("total water discharge by quality and destination") EN22 ("total weight of waste by type and disposal method") and MM4 were each reported by 8 MCGs.

The Table 5.3 also depicted that 7 MCGs reported environmental indicators such as EN27 ("percentage of products sold and their packaging materials that are reclaimed by category"), EN28 ("Monetary value of significant fines and total number of non-monetary sanctions for noncompliance with environmental laws and regulations"), EN29 ("significant environmental impacts of transporting products and other goods and materials used for the organization's operations, and transporting members of the workforce") and two MM3 and MM6. Whilst EN4 ("indirect energy consumption by primary source"), EN7 ("initiatives to reduce indirect energy consumption and reductions achieved"), EN23 ("Total number and volume of significant spills") and MM5 were reported by 6 MCGs, 5 MCGs also reported EN8 ("Total water withdrawal by source"), EN15 (Number of IUCN Red List species and national conservation list species with habitats in areas affected by operations, by level of extinction risk"), EN18 ("Initiatives to reduce greenhouse gas emissions and reductions achieved") and EN30 ("total environmental protection expenditures and investments by type").

Finally, 4 MCGs disclosed indicators type EN6 ("Initiatives to provide energy-efficient or renewable energy-based products and services, and reductions in energy requirements as a result of these initiatives"), EN9 ("Water sources significantly affected by withdrawal of water"), EN10 ("Percentage and total volume of water recycled and reused") and EN25 ("Identity, size, protected status, and biodiversity value of water bodies and related habitats significantly affected by the reporting organization's discharges of water and runoff"). Three MCGs also reported EN5 ("Energy saved due to conservation and efficiency improvements") and EN14 ("Strategies, current actions, and future plans for managing impacts on biodiversity").

Study conducted by Alazzani and Wan-Hussin (2013) found that the following three environmentally responsible practices were implemented by all but one of the eight sampled companies: EN13, Habitats protected or restored; EN18, Initiatives to reduce greenhouse gas emissions and reductions achieved; and EN23, Total number and volume of significant spills. According to the study, no company disclosed information about: EN27, Percentage of products sold and their packaging materials that are reclaimed by category, due to the fact that this requirement is not applicable to them to oil and gas companies which are the sampled companies.

Table 5.3 Disclosures of Environmental performance indicators

	AGA		CHI		GSR		NGL		TGF		GMC		GBL		EML		NMR		AMR		Total	
	08	12	08	12	08	12	08	12	08	12	08	12	08	12	08	12	08	12	08	12	08	12
EN1	1	1	0	1	1	1	0	1	1	1	1	1	0	1	1	1	0	1	1	1	6	10
EN2	0	1	0	1	1	1	1	1	0	1	0	1	0	1	0	1	1	1	1	1	4	10
EN3	1	1	1	1	0	1	1	1	0	1	0	1	1	1	1	1	1	1	0	1	6	10
EN4	1	1	0	1	1	1	0	1	0	1	0	1	0	0	0	0	0	0	0	0	2	6
EN5	0	0	0	1	0	0	0	1	0	1	0	0	0	0	0	0	0	0	0	0	0	3
EN6	0	0	1	1	0	0	0	1	0	0	0	1	0	0	0	0	0	0	0	1	1	4
EN7	0	0	0	1	1	1	1	1	0	1	0	1	0	0	0	0	0	0	0	1	2	6
EN8	0	0	1	1	1	1	1	1	0	1	0	1	0	0	0	0	0	0	0	0	3	5
EN9	0	0	0	1	0	1	0	1	0	1	1	1	0	0	0	0	0	0	0	0	1	4
EN10	0	0	0	1	0	1	1	1	0	1	0	1	0	0	0	0	0	0	0	0	1	4
EN11	1	1	1	1	0	1	0	1	0	1	0	1	1	1	0	1	0	1	0	0	3	9
EN12	1	1	1	1	1	1	1	1	0	1	0	1	0	1	0	1	0	1	0	0	4	9
EN13	1	0	0	1	1	1	1	1	0	1	0	1	0	0	0	0	0	0	0	1	3	6
EN14	0	0	0	1	0	0	0	1	0	1	0	0	0	0	0	0	0	0	0	0	0	3

	EN15	EN16	EN17	EN18	EN19	EN20	EN21	EN22	EN23	EN24	EN25	EN26	EN27	EN28	EN29	EN30	MM3
	5	10	10	5	8	8	8	8	6	2	4	8	7	7	7	5	7
	2	7	2	2	2	1	4	7	3	0	1	4	5	4	4	3	0
	1	1	1	1	1	0	1	1	0	0	1	1	1	1	0	0	1
	0	1	0	0	1	0	0	1	0	0	0	0	1	0	0	0	0
	1	1	1	0	1	1	0	0	0	0	0	1	1	1	1	0	0
	0	1	1	1	1	0	1	0	1	0	0	0	0	0	0	0	0
	0	1	1	1	1	0	1	0	0	0	1	0	0	1	1	1	1
	0	0	1	1	0	0	0	1	0	0	0	0	0	1	1	0	0
	0	1	1	0	1	1	1	1	1	0	0	1	1	1	0	0	1
	0	1	0	0	0	0	1	0	0	0	0	0	1	0	0	0	0
	1	1	1	1	1	1	1	1	1	1	0	0	1	1	1	1	1
	0	1	0	0	0	0	0	1	0	0	0	1	1	0	0	0	0
	1	1	1	1	1	1	1	1	1	1	0	0	1	1	1	1	1
	0	1	0	0	0	0	1	0	1	0	0	1	0	0	0	1	0
	1	1	1	1	1	1	1	1	1	1	1	1	1	1	1	1	1
	1	1	1	0	1	0	1	0	1	0	0	1	1	0	1	0	0
	0	1	1	0	0	1	0	1	0	0	0	1	0	0	1	0	0
	0	1	1	0	0	1	0	1	1	0	0	0	0	0	1	0	0
	1	1	1	1	1	1	1	1	1	0	0	1	1	0	1	1	1
	1	0	0	1	1	1	0	0	1	0	0	0	1	1	1	1	0
	0	1	1	0	1	1	1	1	1	0	0	1	1	1	0	0	1
	0	0	0	0	0	1	0	1	1	0	0	0	0	1	1	1	0

MM4	0	1	0	1	0	1	0	1	0	1	0	0	0	1	0	8
MM5	0	1	0	0	0	1	0	1	0	1	0	0	0	0	1	0	6
MM6	0	1	0	1	0	0	0	1	0	1	0	0	0	1	0	7
EnpiDI	12	20	14	30	10	16	17	34	31	6	31	5	21	7	20	4	13	6	24	87	225

Source: Author

Notes: AGA = AngloGold Ashanti, CHI = Chirano, GSR = Golden Star Resources, NGL = Newmont, TGF = Goldfields, GMC = Ghana Manganese Company, GBL = Bauxite, EML = Endeavour, NMR = Noble, AMR = Persues

5.5 Human Rights Indicators (HR)

According to GRI the human rights performance indicators aspects comprises of Investment and Procurement Practices, Non-discrimination, Freedom of Association and Collective, Bargaining, Child Labour, Forced and Compulsory Labour, Security Practices, Indigenous Rights, Assessment and Remediation. It is expected the companies report in relation to these aspects (GRI, 2010). The disclosure indices for the human rights in the SR is 0.582, that is, fifty eight percent of items pertaining to human rights indicators were disclosed in the sustainability reports by the MCGs in 2012. From the Table 5.3 HR1 ("percentage and total number of significant investment agreements that include human rights clauses or that underwent human rights screening"), as well as on HR2 ("percentage of significant suppliers and contractors that have undergone screening on human rights and actions taken") were the only two indicators that were reported by all the MCGs with the period.

Table 5.3 also shows that eight MCGs reported HR6 ("Operations and significant suppliers identified as having significant risk for incidents of child labour, and measures taken to contribute to the effective abolition of child labour") and HR7 ("Operations and significant suppliers identified as having significant risk for incidents of forced or compulsory labour, and measures to contribute to the elimination of all forms of forced or compulsory labour"). Seven six and five MCGs made mention of both HR4 ("total number of incidents of discrimination and actions taken"), HR5 ("operations identified in which the right to exercise freedom of association or collective bargaining may be at significant risk, and actions taken to support these rights") and HR8 ("percentage of security personnel trained in the organization's polices or procedures concerning aspects of human rights that are relevant to operations") respectively.

Table 5.4 Disclosures of Human rights performance indicators

	AGA		CHI		GSR		NGL		TGF		GMC		GBL		EML		NMR		AMR		Total	
	08	12	08	12	08	12	08	12	08	12	08	12	08	12	08	12	08	12	08	12	08	12
HR1	1	1	1	1	1	1	1	1	1	1	1	1	0	1	0	1	1	1	0	1	7	10
HR2	1	1	1	1	1	1	0	1	1	1	0	1	1	1	0	1	0	1	0	1	5	10
HR3	0	0	0	1	0	0	1	1	0	1	0	0	0	0	0	1	0	0	0	0	1	4
HR4	0	1	1	1	0	1	0	1	0	1	1	1	0	1	0	0	0	0	0	1	2	7
HR5	0	1	0	1	0	0	1	1	0	1	0	0	0	1	0	1	0	0	0	0	1	6
HR6	0	1	0	1	0	1	0	1	1	1	0	1	0	1	0	0	0	1	0	0	1	8
HR7	0	1	0	1	0	1	0	1	1	1	0	1	0	1	0	0	0	1	0	0	1	8
HR8	0	0	0	1	0	0	0	1	0	1	0	0	0	0	0	0	0	0	0	0	0	3
HR9	0	0	0	0	0	1	0	1	0	1	0	0	0	0	0	0	0	0	0	1	0	4
HR10	0	0	0	0	0	0	0	0	0	1	0	0	0	0	0	0	0	1	0	0	0	2
HR11	0	0	0	0	0	0	0	0	1	1	0	0	0	0	0	0	0	1	0	1	1	2
HRPDI	2	6	3	8	2	6	3	9	5	11	2	5	1	6	0	4	1	5	0	4	19	64

Source: Author

Notes: AGA = AngloGold Ashanti, CHI = Chirano, GSR = Golden Star Resources, NGL=Newmont, TGF =Goldfields, GMC = Ghana Manganese Company, GBL = Bauxite, EML = Endeavour, NMR = Noble, AMR = Persues

Another indicators that were fairly reported are HR3 ("Total hours of employee training on policies and procedures concerning aspects of human rights that are relevant to operations, including the percentage of employees trained") and HR9 ("Total number of incidents of violations involving rights of indigenous people and actions taken"). Four MCGs were known according to the Table to have report on them in their SR within the period. Finally 2 MCGs' SR was found to have reported HR10 ("Percentage and total number of operations that have been subject to human rights reviews and/or impact assessments") and HR11 ("Number of grievances related to human rights filed, addressed, and resolved through formal grievance mechanisms").

5.6 Labour Practices & Decent Work Performance Indicators

Labour practices & decent work performance indicators (LA) looks at the broad issues on Employment, Labour/Management Relations, Occupational Health and Safety, Training and Education, Diversity and Equal Opportunity and Equal Remuneration for Women and Men (GRI 2010). Table 5.5 shows that all MCGs disclosed items such as employee compensation, welfare and donation, executive profile, in-house training arrangement for the employees, and appreciating and motivating employees for their efforts in SR. From the total of 170 items relating to LA disclosure, 94 items were disclosed in the SRs by MCGs. The labour practices & decent work performance indicator disclosure indices for MCGs indicate that 55 percent items of LA issues were disclosed in the sustainability reports.

It can be seen from Table 5.5 that at least there hasn't been one indicator that has not been reported on by MCGs. About nine MCGs were seen to have reported on LA2 ("total number and rate of employee turnover by age group, gender, and region") and LA10 ("average hours of training per year per employee by employee category"). Eight MCGs reported on LA1 ("Total workforce by employment type, employment contract, and region, broken down by gender"), LA7 ("rates of injury, occupational diseases, lost days, absenteeism and total number of work-related fatalities, by region") and LA14 ("Ratio of basic salary and remuneration of women to men by employee category, by significant locations of operation"). Seven MCGs reported on LA4 ("percentage of employees covered by collective bargaining agreements"), LA5 and LA8 ("Education, training, counselling, prevention, and risk-control programs in place to assist workforce members, their families, or community members regarding serious diseases"). Six MCGs also reported on LA11 ("Programs for skills management and lifelong learning that support the continued employability of employees and assist them in managing career endings") and LA13 ("Composition of governance bodies and breakdown of employees per employee category according to gender, age group, minority group membership, and other indicators of diversity").

Table 5.5 Disclosures of Labour practices & decent work performance indicators

	AGA		CHI		GSR		NGL		TGF		GMC		GBL		EML		NMR		AMR		Total	
	08	12	08	12	08	12	08	12	08	12	08	12	08	12	08	12	08	12	08	12	08	12
LA1	1	1	0	1	1	1	1	1	1	1	1	1	0	1	0	0	1	1	0	0	6	8
LA2	0	1	1	1	0	1	0	1	1	1	0	1	1	1	1	1	0	1	1	0	5	9
LA3	0	0	0	1	0	0	0	1	1	1	0	1	0	0	0	0	0	0	0	1	1	5
LA4	0	1	0	1	0	0	1	1	0	1	1	1	1	1	0	0	0	0	0	1	3	7
LA5	0	1	0	1	0	0	0	1	0	1	0	1	1	1	0	0	0	0	0	0	2	7
LA6	0	0	0	1	0	0	0	1	0	1	0	1	0	0	0	1	0	1	0	0	0	5
LA7	1	1	0	1	0	1	0	1	1	1	0	1	0	1	0	1	0	0	0	0	2	8
LA8	1	1	0	1	0	0	0	1	1	1	0	1	0	1	0	0	0	1	0	0	2	7
LA9	0	0	0	1	0	0	0	1	0	1	0	0	0	0	0	1	0	1	0	0	0	5
LA10	0	1	1	1	0	1	1	1	0	1	0	1	0	1	0	0	0	1	0	1	2	9
LA11	0	0	1	1	0	1	0	1	0	1	0	0	0	0	0	0	0	1	0	1	1	6
LA12	0	0	1	1	0	0	0	1	1	1	0	0	0	0	0	0	0	0	0	0	2	3
LA13	0	1	0	1	0	0	0	1	0	1	0	1	0	1	1	0	0	0	0	1	0	6
LA14	1	1	1	1	0	0	1	1	1	1	0	0	0	1	0	0	0	0	0	1	4	8
LA15	0	0	0	0	0	0	0	0	0	0	0	0	0	0	0	0	0	0	0	0	0	1
MM12	0	0	0	0	0	0	0	0	0		0	0	0	0	0	0	0	0	0	0	0	0

MM13	0	0	0	0	0	0	0	0	0	0	0	0	0	0	0	0	0	0	0	0	0
LAPDI	**4**	**9**	**5**	**14**	**1**	**5**	**14**	**7**	**14**	**2**	**10**	**3**	**9**	**1**	**5**	**1**	**7**	**1**	**6**	**30**	**94**

Source: Author

Notes: AGA = AngloGold Ashanti, CHI = Chirano, GSR = Golden Star Resources, NGL = Newmont, TGF = Goldfields, GMC = Ghana Manganese Company, GBL = Bauxite, EML = Endeavour, NMR = Noble, AMR = Persues

Furthermore five MCGs also reported on LA3 ("Benefits provided to full-time employees that are not provided to temporary or part-time employees, by significant locations of operation"), LA6 ("percentage of total workforce represented in formal joint management-worker health and safety committees that help monitor and advise on occupational health and safety programs") and LA9 ("Health and safety topics covered in formal agreements with trade unions"). Three and two of MCGs also reported on LA12 ("Percentage of employees receiving regular performance and career development reviews, by gender") and LA15 ("Return to work and retention rates after parental leave, by gender") respectively.

5.7 Product Responsibility Performance Indicators (PR)

The product responsibility performance indicators (PR) comprises of Customer Health and Safety, Product and Service Labelling, Marketing Communications, Customer Privacy, and Compliance (GRI, 2010). Out of total of 11 items relating to product responsibility performance indicators, 8 items were disclosed in the SR of the MCGs (see Table 5.6). The disclosure indices for product responsibility performance indicators disclosure show that 46 percent of items concerning product responsibility issues were disclosed in the annual reports in 2012. Table 5.6 shows the product responsibility indicators for the MCGs in the 2012 shows that all the MCGs fail to report every indicator within the period under consideration. Nine MCGs reported PR1 ("life cycle stages in which health and safety impacts of products and services are assessed for improvement, and percentage of significant products and services categories subject to such procedures") in their SRs.

Table 5.6 Disclosures of G3 Product responsibility performance indicators

	AGA		CHI		GSR		NGL		TGF		GMC		GBL		EML		NMR		AMR		Total	
	08	12	08	12	08	12	08	12	08	12	08	12	08	12	08	12	08	12	08	12	08	12
PR1	1	1	1	1	1	1	0	1	0	0	1	1	1	1	1	1	1	1	1	1	7	9
PR2	0	1	1	1	0	1	1	1	0	0	0	0	0	1	1	1	0	1	0	1	3	7
PR3	1	1	1	1	0	0	0	1	0	0	0	0	1	1	1	1	1	1	0	1	4	7
PR4	0	0	0	1	0	0	1	1	0	0	0	0	0	0	0	0	0	1	0	0	1	3
PR5	0	0	0	1	0	0	0	1	0	0	0	0	0	0	0	0	0	0	0	0	0	2
PR6	0	1	0	1	0	1	0	1	0	0	0	1	1	1	0	1	0	0	1	1	1	2
PR7	0	0	0	1	0	0	0	1	0	0	0	0	0	0	0	1	0	1	1	1	1	4
PR8	0	0	0	1	1	1	0	1	0	0	0	0	0	1	1	0	0	1	0	1	2	5
PR9	0	1	1	1	1	0	0	1	0	1	0	0	1	1	0	1	0	0	0	0	2	7
PR10	0	0	0	0	0	0	0	0	0	0	0	0	0	0	0	0	0	0	0	0	0	0
PRPDI	2	5	4	9	3	5	2	9	0	1	1	2	4	4	4	6	2	5	3	6	21	46

Source: *Author*

Notes: AGA = AngloGold Ashanti, CHI = Chirano, GSR = Golden Star Resources, NGL = Newmont, TGF = Goldfields, GMC = Ghana Manganese Company, GBL = Bauxite, EML = Endeavour, NMR = Noble, AMR = Persue

Eight MCGs had their SR included an indicator PR6 ("Programs for adherence to laws, standards, and voluntary codes related to marketing communications, including advertising, promotion, and sponsorship"). Seven MCGs reported on PR2 ("total number of incidents of noncompliance with regulations and voluntary codes concerning health and safety impacts of products and services, by type of outcomes"), PR3 ("type of product and service information required by procedures and percentage of significant products and services subject to such information requirements") and PR9 ("Monetary value of significant fines for non-compliance with laws and regulations concerning the provision and use of products and services"). Three MCGs also had PR4 ("total number of incidents of noncompliance with regulations and voluntary codes concerning product and service information and labelling, by type of outcomes") in their SR. Four MCGs did report on PR7 ("total number of incidents of noncompliance with regulations and voluntary codes concerning marketing communications, including advertising, promotion, and sponsorship, by type of outcomes"). PR5 ("practices related to customer satisfaction, including results of surveys measuring customer satisfaction") were reported by two MCGs but no single one reported on PR10. Finally, five MCGs did report on PR8 ("Total number of substantiated complaints regarding breaches of customer privacy and losses of customer data").

5.8 Society Performance Indicators (SO)

The society performance indicators (SO) looks at issues such as local communities, corruption, public policy, anti-competitive behaviour and compliance issues (GRI, 2010). Table 5.7 shows that out of 28 items associated with community development, 26 items in the annual reports in 2012 by the MCGs. Most of the MCGs have a separate department for performing CSR activities e.g. sponsoring sports and cultural functions, patronizing general and technical education, poverty alleviation programmes, undertaking low cost housing projects, creating job opportunities for unemployed youth and other community issues are preferred in disclosure, whereas, issues such as anti-corruption measures, and community activities within the corporate vicinities are totally ignored in the sustainability reports. The society performance indicators disclosure indices indicate that 72 percent of items pertaining to society issues were disclosed in the SR by the MCGs in 2012.

According to Table 5.7 all the MCGs reported on indicators SO1 ("Percentage of operations with implemented local community engagement, impact assessments, and development programs") and SO2 ("Percentage and total number of business units analysed for risks related to corruption"). Eight MCGs also reported on SO4 ("actions taken in response to incidents of corruption") and SO8 ("Monetary value of significant fines and total number of non-monetary sanctions for noncompliance with laws and regulations"). SO3 ("Percentage of employees trained in organization's anti-corruption policies and procedures"), MM7 and MM10 were also reported by nine MCGs with seven MCGs also reporting on SO5 ("Public policy positions and participation in public policy development and lobbying"), MM8 and MM11. SO6 ("Total value of financial and in-kind contributions to political parties, politicians, and related institutions by country") was reported six times, SO7 ("Total number of legal actions for anticompetitive behaviour, anti-trust, and monopoly practices & their outcomes") reported five times, SO10 ("Prevention and mitigation measures implemented in operations with significant potential or actual negative impacts on local communities") two times whilst SO9 ("Operations with significant potential or actual negative impacts on local communities") was only reported once by MCG.

Table 5.7 Disclosures of Society performance indicators

	AGA 08	AGA 12	CHI 08	CHI 12	GSR 08	GSR 12	NGL 08	NGL 12	TGF 08	TGF 12	GMC 08	GMC 12	GBL 08	GBL 12	EML 08	EML 12	NMR 08	NMR 12	AMR 08	AMR 12	Total 08	Total 12
SO1	1	1	1	1	0	1	1	1	1	1	1	1	1	1	1	1	1	1	1	1	9	10
SO2	1	1	1	1	1	1	1	1	1	1	1	1	1	1	1	1	1	1	1	1	10	10
SO3	0	1	0	1	0	0	1	1	1	1	1	1	1	1	0	1	1	1	1	1	6	9
SO4	1	1	0	0	0	0	0	0	0	0	0	0	0	0	0	0	1	1	0	0	2	2
SO5	0	1	1	1	1	1	1	1	0	1	0	0	1	1	0	0	0	0	0	1	4	7
SO6	0	0	0	0	0	0	0	0	1	1	0	0	0	0	0	0	0	0	0	0	1	1
SO7	0	1	1	1	0	0	1	1	0	1	0	0	0	0	0	0	1	1	0	0	3	5
SO8	1	1	0	0	0	0	0	1	1	1	0	1	1	1	1	1	0	1	0	1	3	8
SO9	0	1	0	1	1	1	0	1	1	1	0	0	0	0	1	0	0	1	0	0	3	6
SO10	1	1	0	1	0	1	0	1	1	1	0	0	0	1	0	1	0	1	0	0	2	8
MM7	0	1	0	1	0	1	0	1	0	1	0	1	0	1	0	1	0	0	0	1	0	9
MM8	0	1	0	0	0	0	0	1	0	1	0	1	0	1	0	1	0	1	0	0	0	7
MM9	0	1	0	1	0	1	0	1	0	1	0	1	0	1	0	1	0	1	0	1	0	10
MM10	0	1	0	1	0	1	0	1	0	1	0	1	0	1	0	0	0	1	0	1	0	9
MM11	0	1	0	1	0	1	0	1	0	1	0	1	0	1	0	0	0	0	0	0	0	7
SOPDI	5	14	4	11	3	9	5	13	7	15	3	9	5	11	4	8	5	11	3	8	43	108

Source: Author

Notes: AGA = AngloGold Ashanti, CHI = Chirano, GSR = Golden Star Resources, NGL=Newmont, TGF =Goldfields, GMC = Ghana Manganese Company, GBL = Bauxite, EML = Endeavour, NMR = Noble, AMR = Persues

5.9 Change in Sustainability Performance Disclosure between 2008 and 2012

According to GRI (2010), social dimension of sustainability is related to the impact of an organization on the social systems within which it operates. In order to minimise issues associated with content analysis such as counting of words or sentences and how to deal with charts and pictures, in consistent with prior studies (Hossain and Adams, 1995; Hossain et al., 1995; Dixon et al., 1994; Barako, et al., 2006), this study uses a disclosure index to reveal the amount of performance indicators disclosed by MCGs (Barako, et al., 2006). A disclosure index involves the researcher identifying whether an entity does or does not disclose an item or indicator in the GRI guideline list (Barako, et al., 2006).

The indices considered are Corporate Sustainability performance indicator disclosure index (CSPDI), made up of (i) Economic performance indicator disclosure index (ECPDI), (ii) Environmental performance indicator disclosure index (ENPDI), (iii) Human right performance indicator disclosure index (HRPDI), (iv) Labour practice & decent work performance indicator index (LAPDI), (v) Product responsibility performance indicator index (PRPDI) and (vi) Society performance indicator disclosure index (SOPDI). The last four categories belong to Social performance indicator index. These are clearly shown in Table 7.9 and these disclosure indices are calculated by dividing the number of items disclosed by the number of items on the score-sheet. The total items were disclosed by each disclosure theme as designated in the research instrument were identified. Mean disclosures were calculated as the number of items disclosed divided by the number of MCGs.

It can be seen from Table 5.8 that out of 980 GRI performance indicators 226 and 625 (with disclosure index of 0.231 and 0.638 respectively) were disclosed in 2008 and 2012 respectively. This means the disclosure within the five years there have increased by 2.77 and this is consistent with the study by KPMG (2008). Almost all the indicators had a substantive increase with the HRPDI and PRPDI being the highest and the lowest respectively with an increase of 6.4 and 2.19 respectively within the period. The reason might be attributed to the fact that there is much awareness of human rights issues as a result of improved communication systems, for example the vibrant radio stations in the country. Table 5.8 also shows that all MCGs disclosed substantial sustainability information in both periods in the annual reports with environmental information being more (39% and 37% in 2008 and 2012 respectively). This is attributed to the fact that mining companies are expected to disclose more information due to the nature of their activities. Interestingly social performance indicators represent 45% and 49% for 2008 and 2012 with only 16% and 14% being economic performance indicators for the same periods.

Table 5.8 Descriptive Statistics for Sustainability Disclosure of MCGs

Themes of Disclosure	GRI Indicators	Indicators Disclosed		% Change from 2008 to 2012	Disclosure Index		Mean of DI		St. Deviation of DI		Median of DI	
		2008	2012		2008	2012	2008	2012	2008	2012	2008	2012
ECPDI	110	35	88	2.51	0.318	0.80	3.5	8.0	1.27	2.280	3.5	9
ENPDI	340	87	225	2.59	0.256	0.662	8.7	6.628	4.35	2.257	6.5	22.5
HRPDI	110	10	64	6.4	0.090	0.582	1.8	5.529	1.47	2.993	1.5	6
LAPDI	170	30	94	3.13	0.176	0.553	3	4.6	2.05	2.939	2.5	9
PRPDI	100	21	46	2.19	0.21	0.46	2.1	4.6	1.45	2.875	2	5
SOPDI	150	43	108	2.51	0.287	0.72	4.4	7.2	1.26	2.757	4.5	11
CSPDI	980	226	625	2.77	0.231	0.638						

Source: Author

CSPDI = Corporate Sustainability performance indicator disclosure index, ECPDI = Economic performance indicator disclosure index, ENPDI = Environmental performance indicator disclosure index, HRPDI= Human right performance indicator disclosure index, LAPDI=Labour practice & decent work performance indicator index, PRPDI= Product responsibility performance indicator index and SOPDI =Society performance indicator disclosure index.

Table 5.9 presents the company level analysis in terms of trends of the MCGs sustainability performance indicators disclosure for 2008 and 2012. It also presents the percentage change of disclosure within the two periods under consideration. The table shows that the number of performance indicators disclosure has more than double with respect of all the MCGs (2.10, 2.34, 2.23, 2.43, 2.28, 4.71, 3.21, 2.78, 3.27 and 3.31). These further confirm the fact that there has been an increment of transparency as a result of the fact that many stakeholders are now expecting companies to provide more information in their SR. It can also be deduced from Table 5.9 that companies such as GMC, GBL, NMR and AMR with disclosure increases between 3.21 and 4.71. The main reason for these increases in disclosure in SR of these companies is that almost all of these companies were either started operation with the last ten years or acquired within the period.

The analysis of the results indicates that the extent of performance indicators reported in sustainability reports of the MCGs is consistent with study by Gurvitsh and Sidorova (2012). All of the three areas in the triple bottom line were widely addressed to give credence to support the definitions of sustainability reporting, although the balance varied considerably. Most of the indicators suggested by the GRI G3 guidelines were reported by all the MCGs and this confirmed the findings by Brown et al. (2009a) that the GRI is becoming an established institution, in spite of the short time since its existence. Notwithstanding, there were some remarkable differences in the frequency of reporting the environmental, economic, and social indicators suggested by the GRI G3 guidelines.

Table 5.9 Change in MCGs performance indicators disclosure between 2008 and 2012

	AGA		CHI		GSR		NGL		TGF		GMC		GBL		EML		NMR		AMR	
	08	12	08	12	08	12	08	12	08	12	08	12	08	12	08	12	08	12	08	12
ECPDI	5	9	5	10	3	8	5	11	4	11	2	9	4	10	2	7	2	8	3	5
ENPDI	12	20	14	30	10	16	17	34	6	31	6	31	5	21	7	20	4	13	6	24
HRPDI	2	6	3	8	2	6	3	9	5	11	1	5	1	6	0	4	1	5	0	4
LAPDI	4	9	5	14	1	5	5	14	7	14	2	10	3	9	1	5	1	7	1	6
PRPDI	2	5	4	9	3	5	2	9	0	1	0	2	1	4	4	6	2	5	3	6
SOPDI	5	14	4	11	3	9	5	13	7	15	3	9	5	11	4	8	5	11	3	8
CSPDI	30	63	35	82	22	49	37	90	32	73	14	66	19	61	18	50	15	49	16	53
% Change	2.10		2.34		2.23		2.43		2.28		4.71		3.21		2.78		3.27		3.31	

Source: Author

Notes:

1. AGA= AngloGold Ashanti, CHI= Chirano, GSR= Golden Star, NGL=Newmont, TGF=Goldfields, GMC= Ghana Manganese Company, GBL= Bauxite, EML= Endeavour, NMR= Noble, AMR= Persues
2. CSPDI = Corporate Sustainability performance indicator disclosure index, ECPDI = Economic performance indicator disclosure index, ENPDI= Environmental performance indicator disclosure index, HRPDI= Human right performance indicator disclosure index, LAPDI=Labour practice & decent work performance indicator index, PRPDI= Product responsibility performance indicator index and SOPDI =Society performance indicator disclosure index.

The result is consistent with previous research made by Skouloudis and Evangelinos (2009) and Gallego (2006) that there are similarities and differences of performance indicators reported by the MCGs. For example, economic indicators focusing on sales and benefits were widely reported in all studies. Environmental indicators focusing on energy and water were also widely reported. With reference to social indicators, there were also several similarities including indicators focused on donations, labour practices, and the breakdown of the workforce. However, there were variations of indicators reported and their frequency of occurrence did vary considerably.

Some of the possible reasons for the differences in disclosure include sample size, sample composition, and the fact that only one sector was involved (Roca and Searcy, 2012). Another reason for the differences is the fact that different countries have different national business systems that are shaped by a variety of national institutions (Matten and Moon, 2008). Again, the Institutional factors such as political systems, financial systems, education and labour systems, cultural customs, the organization of market processes and corporate governance have a lot of implications for the structure of corporations (Matten and Moon, 2008; Roca and Searcy, 2012). The evidence noted above is that the analysis of corporate sustainability reports showed an incredible variety in the indicators among the MCGs. There are a number of possible explanations for this variety that may be drawn from the nature of the sample, the previous literature and the theoretical frameworks used to explore corporate sustainability. For example, reports explicitly focused on the environment would be expected to focus more on environmental, rather than economic or social issues.

It was also realised that the length of the reports varied widely and these differences are as a result of the lack of agreement on the information MCSs are expected to disclose. This varied nature of report length contributed to the different scopes and therefore different performance indicators reported (Roca and Searcy, 2012). These variations may also be explained by the differing purposes and target audiences of the SRs. Although relatively little detail on the purpose and intended audiences of the reports was provided by the companies in the sample, it is clear that these varied widely as well.

The studies on SR also provide insight into the diversity observed in the PIs. According to Roca and Searcy (2012), there are differing interpretations of terms such as "sustainability" and "corporate social responsibility" and this has contributed to differences in performance indicators reported in SRs. Furthermore, lack of standardization of SRs has been seen and many research works have revealed the wide latitude provided by the GRI can lead to differences in indicator disclosure (Moneva et al. 2006). Apart from the mining sector, many other sectors of the economy have been previously studied (Ratanajongkol et al., 2006; Cooke, 1989). Studies made by Deegan and Gordon (1996) alluded that companies whose main operations have serious impact on the environment are expected to disclose more information on social responsibility than those whose activities have less impact on the environment. In another development, several researchers have noted that corporate sustainability initiatives and approaches must be made in relation to the circumstances of the locality (Van Marrewijk, 2003; Steurer et al., 2005).

There have been many theoretical explanations given to the wide variety of indicators disclosed by companies. Two relevant perspectives are provided by stakeholder theory and legitimacy theory; although other theories can be used to explain the perspective. As explained in the previous Chapter, stakeholder theory holds that corporations have obligations to a number of individuals and groups called stakeholders. It is also known that different corporations have different stakeholders. Here, if different corporations have different priorities for different stakeholders, it can therefore be assumed that SR, which is targeted to those stakeholders, would disclose different indicators especially in the absence of mandatory reporting requirements. Deegan et al. (2000) argued that all stakeholders should be treated equally regardless of their relative power but others researchers are of the view that the power of a corporation's stakeholders are continuously changing (Mitchell et al., 1997).

The legitimacy theory also state that corporations are a part of larger society and must operate within the bounds set by that society (Suchman, 1995). In view of this theory, SR may be viewed as a part of strategy of organisations to build and maintain its legitimacy (Ratanajongkol et al., 2006). Furthermore, due to the fact that different corporations are subject to different expectations from society, they may find it necessary to report different performance indicators in their SR so as to be perceived as legitimate. There may be other reasons for the diversity of indicators disclosed but the two key factors that must receive particular attention are as follows. One of the reasons that may be attributed to the diversity of PIs disclosure is the fact that there were few mandatory requirements for sustainability reporting in Ghana. The MCGs are not required to report on specific indicators in their SRs. Secondly, in spite of the many research work done, sustainability reporting practices is still in its relatively early stages. Again the voluntary sustainability reporting guidelines as well as the best practices by industry leaders provide a starting point but there is still considerable discretion in the selection of what information to be made available to the public.

Statistics from GRI also reflect a growing uptrend in the world that more and more companies are using the Global Reporting Initiative's (GRI) to measure and report their economic, environmental, social and governance performance which have annually increase from 22 to 58 per cent within the periods of 2006 to 2011 (Gurvitsh and Sidorova, 2012). This development is as a result of new audiences for sustainability information, like investors and regulators, who are now calling for more and better performance data and these annual growths in the number of reporters is expected to continue (GRI, 2010; Gurvitsh and Sidorova, 2012). The study is also consistent with the study by KPMG in 2008 which shown that corporate social reporting has become the mainstream and a clear uptrend trend in reporting, whereby 80% of the largest 250 companies worldwide issued such reports, as opposed to 50% in the 2005 survey (KPMG, 2008). KPMG study also found that CSR is now not the exception of the world's largest companies with climate change reporting on the increase though improvement is expected (KPMG, 2008; Khalid et al., 2013). Finally this study also revealed that MCGs prefer to include sustainability disclosures in their annual reports rather than issue stand-alone sustainability or CSR reports which is consistent with the study by Gurvitsh and Sidorova (2012).

5.10 Summary

In conclusion the main objective of this chapter was to identify the disclosure of corporate sustainability performance indicator in the SRs of MCGs in Ghana. It was found that all the MCGs disclosed at least one item of sustainability information in the SR which is consistent with Sobhani et al. (2012) research. Social performance indicators disclosures received more attention than environmental sustainability issues in disclosure despite of the fact that they are mining companies (see also Belal, 2007). The findings are consistent with the recent studies of corporate social and environmental disclosure Sobhani et al. (2009), in the banking industry where it was revealed that more social and less environmental information are being disclosed by companies. Earlier studies showed that social and environmental information was rarely disclosed under separate headings (see Belal, 2001) but consistent with study by Sobhani et al. (2012), this study found remarkable development with respect to the location of disclosure and its presentation styles under the SR heading. This is in conformity with the GRI sustainability guidelines.

Furthermore, content analysis of the sustainability reports of the ten mining companies in Ghana indicates that they made reasonable efforts to disclose their environmental performance in accordance with the GRI guidelines which appear to provide a robust and readily available tool for reporting comprehensive progress concerning all aspects of social, economic and environmental activities. The results PIs related to the environmental dimension of sustainability were represented in higher proportion than the economic but smaller in social dimensions by the MCGs. This is despite the fact that the study is on the mining companies which are known to affect the environment on a massive scale. The voluntary adoption of the guidelines by a vast majority of companies increases transparency, credibility and comparability in sustainability reporting. It is recommended that Government of Ghana and professional bodies should support the adoption of these international reporting standards to add value or credibility to the reports of those companies that have adopted it.

REFERENCES

Adams, C.A. and Frost, G.R. (2008). Integrating sustainability reporting into management practices. **Accounting Forum**, 32 pp. 288-302.

Adhikari, A. and Tondkar, R.H., (1992). Environmental factors influencing accounting disclosure requirements of global stock exchanges. **Journal of International Financial Management & Accounting**, *4*(2), pp.75-105.

Al-Akra, M., Ali, M.J. and Marashdeh, O. (2009). Development of accounting regulation in Jordan. **The International Journal of Accounting**, *44*(2), pp.163-186.

Alazzani, A. and Wan-Hussin, W.N. (2013). Global Reporting Initiative's environmental reporting: A study of oil and gas companies. **Ecological indicators**, *32*, pp.19-24.

Ali, M. J. and Ahmed,K. (2007). The legal and institutional framework for corporate financial reporting practices in South Asia. **Research in Accounting Regulation**, 19 (2007), pp. 175–205

American Institute of Certified Public Accountants (AICPA). (1994) **Improving Business Reporting**. A Customer Focus (New York: AICPA).

Aryee, B.N. (2001). Ghana's mining sector: its contribution to the national economy. **Resources Policy**, *27*(2), pp.61-75.

Ashraf, J.,and Ghani, W. I. (2005). Accounting development in Pakistan. The International Journal of Accounting, 40(2), pp. 175-201.

Assenso-Okofo, O., Ali, M.J. and Ahmed, K., (2011). The development of accounting and reporting in Ghana. **The International Journal of Accounting**, *46*(4), pp.459-480

Azapagic, A. (2004). Developing a framework for sustainable development indicators for the mining and minerals industry. **Journal of Cleaner Production** 12 (6) pp. 639-662.

Ball, R., Robin, A. and Wu, J.S. (2003). Incentives versus standards: properties of accounting income in four East Asian countries. *Journal of accounting and economics*, *36*(1-3), pp.235-270.

Barako, D. G., Hancock, P. and Izan, H. Y. (2006). Factors Influencing Voluntary Corporate Disclosure by Kenyan Companies. **Corporate Governance**, 14 (2) March, pp. 107-125.

Bebbington, J., Larrinaga, C. and Moneva, J. M. (2008). Corporate Social Reporting and Reputation Risk Management. **Accounting, Auditing and Accountability Journal**, 21 (3) pp. 337–361.

Beck, T., Demirgüç-Kunt, A. and Levine, R., (2003). Law and finance: why does legal origin matter? **Journal of comparative economics**, *31*(4), pp.653-675.

Beloe, S., Elkington, J., Hester, K.F., Loose, M. and Zollinger, P. (2006). Tomorrow's Value: The Global Reporters 2006 Survey of Corporate Sustainability Reporting. SustainAbility, London.

Berry, L.B. ed., (1995). **Ghana: A country study** (Vol. 550, No. 153). US Government Printing Office.

Boateng, E.A. (1967). **Politics and education** (No. 1). Liberty Press.

Boersema, J.M. and Van Weelden, S.J. (1992). **Financial Reporting for Segments** (Toronto: Canadian Institute of Chartered Accountants).

Bouten, L., Everaet, P. and Roberts, P. W. (2012). How a Two-Step Approach Discloses Different Determinants of Voluntary Social and Environmental Reporting. **Journal of Business Finance & Accounting**, 39(5) & (6), pp. 567–605.

Bouten, L., P., Everaert, L., Van Liedekerke, L., De Moor and Christiaens, J. (2011) Corporate Social Responsibility Reporting: A Comprehensive Picture. **Accounting Forum**, 35(3) (September), pp. 187–204.

Brown, H.S., de Jong, M., Levy, D.L. (2009a). Building institutions based on information disclosure: lessons from GRI's sustainability reporting. **Journal of Cleaner Production** 17 (6) pp. 571-580.

Brown, J. and Fraser, M. (2006). Approaches and perspectives in social and environmental accounting: an overview of the conceptual landscape. **Business Strategy and the Environment,** 15 pp. 103-117.

Bushman, R.M. and Piotroski, J.D., (2006). Financial reporting incentives for conservative accounting: The influence of legal and political institutions. **Journal of Accounting and Economics,** *42*(1-2), pp.107-148.

Camfferman, K. and Cooke, T.E. (2002). An analysis of disclosure in the annual reports of UK and Dutch companies. **Journal of International Accounting Research**, *1*(1), pp.3-30.

Carroll, A. and Buchholtz, A. (2006). **Business and society: Ethics and stakeholder management**. 6th ed. Mason, Ohio: Thompson South-Western.

Chee Tahir, A. and Darton, R. C. (2010). The process analysis method of selecting indicators to quantify the sustainability performance of a business operation. **Journal of Cleaner Production** 18 pp. 1598-1607.

Clarkson, P. M., Li, Y., Richardson, G. D., and Vasvari, F. P. (2008). Revisiting the relation between environmental performance and environmental disclosure: an empirical analysis. **Accounting, Organizations and Society**, 33 (4/5) pp. 303-327.

Cooke T. E. (1989). Disclosure in the Corporate Annual Reports of Swedish Companies. **Accounting and Business Research**, 74 pp. 13-24.

Cooke T. E. (1991). An assessment of Voluntary Disclosure in the Annual Reports of Japanese Corporations. **International Journal of Accounting**, 26 pp. 174-189.

Cooke T. E. (1992). The Impact of Size, Stock Market Listing and Industry Type on Disclosure in the Annual Reports of Japanese Listed Corporation. **Accounting and Business Research**, 22 pp. 229-237.

CSBR (2008). CSR Trends 2008: Our Second Comprehensive Survey of Sustainability Report Trends, Benchmarks and Best Practices, **Canadian Business for Social Responsibility and Craib Design and Communication**, Toronto.

Daub, C. H. (2007). Assessing the quality of sustainability reporting: an alternative methodological approach. **Journal of Cleaner Production,** 15 pp. 75-85.

Davis, G. and Searcy, C. (2010). A review of Canadian corporate sustainable development reports. **Journal of Global Responsibility** 1 (2) pp. 316-329.

Davis, K. (1973). The case for and against business assumption of social responsibilities. **Academy of Management Journal**, 16, pp. 312-322.

Deegan C. and Gordon, B. (1996). A study of environmental disclosure practices of Australian Corporations. **Accounting and Business Research**, 26 (3) pp. 187-199.

Deegan, C., Rankin, M. and Voght, P. (2000). Firms' disclosure reactions to major social incidents: Australian evidence. **Accounting Forum**, 24 (1) pp. 101-130.

Deegan, C., Rankin, M., and Tobin, J. (2002). An examination of the corporate social and environmental disclosures of BHP from 1983-1997: a test of legitimacy theory. **Accounting, Auditing and Accountability Journal**, 15 (3), pp. 312-343.

Deppe, L. and Omer, S. C. (2000). Disclosing disaggregated information. **Journal of Accountancy,** 190(3), pp. 47–52.

Dixon, R.K., Winjum, J.K., Andrasko, K.J., Lee, J.J. and Schroeder, P.E. (1994). Integrated land-use systems: assessment of promising agroforest and alternative land-use practices to enhance carbon conservation and sequestration. **Climatic change,** 27(1), pp.71-92.

Edwards, P. and Smith, R. A. (1996). Competitive disadvantage and voluntary disclosures: the case of segmental reporting. **British Accounting Review,** 28, pp. 155-172.

Elkington, J. (1998). **Cannibals with Forks: The Triple Bottom Line of 21st Century Business**. New Society Publishers, Stony Creek, CT.

Eng, L.L. and Mak, Y. T. (2003). Corporate governance and voluntary disclosure. **Journal of Accounting and Public Policy,** 22 pp. 325–345

Farneti, F., and Guthrie, J. (2009, June). Sustainability reporting by Australian public sector organisations: Why they report. **In Accounting forum,** 33 (2), pp. 89-98.

Gallardo, J., (2002). **A framework for regulating microfinance institutions: The experience in Ghana and the Philippines**.

Gallego, I. (2006). The use of economic, social and environmental indicators as a measure of sustainable development in Spain. **Corporate Social Responsibility and Environmental Management** 13 pp. 78-97.

Ghana National Accounting Standards, ICA: i-vi CSIR Accra Available at: http://www.icagh.com/ (2000)

Ghana Stock Exchange Fact Book, 2007, Ghana Stock Exchange Fact Book

Ghana Stock Exchange Fact Book, Ghana Stock Exchange Publication, Accra (2009)

Ghana Stock Exchange Publication, Accra (2007), Ghana Stock Exchange Fact Book, 1990

Ghanaweb, 2009 Ghanaweb Available at: http://www.ghanaweb.com/GhanaHomePage/country/ (2009)

Global Reporting Initiative (GRI), (2006). Sustainability Reporting Guidelines Version 3.0. GRI, Amsterdam.

Global Reporting Initiative (GRI), (2010). Sustainability Reporting Guidelines Version 3.0. GRI, Amsterdam.

Global Reporting Initiative (GRI), (2011a). What is GRI? Available from: http://www. globalreporting.org/AboutGRI/WhatIsGRI/ [Accessed June 17, 2011].

Global Reporting Initiative (GRI), (2011b). Sustainability Reporting Guidelines: Version 3.1. GRI, Amsterdam.

Global Reporting Initiative (GRI), (2012). Sustainable Reporting Guidelines G3. Amsterdam.

Global Reporting Initiative (GRI), 2011a. What is GRI? Available from: http://www. globalreporting.org/AboutGRI/WhatIsGRI/ [accessed June 17, 2011].

Global Reporting Initiative (GRI), 2011b. Sustainability Reporting Guidelines: Version 3.1. GRI, Amsterdam.

Global Reporting Initiative (GRI). 2006. Sustainability Reporting Guidelines: Version

Gonella C., Pilling, A. and S. Zadek (1998) Making Values Counts: Contemporary Experience in Social and Ethical **Accounting, Auditing and Reporting**. ACCA Research Report 57.

Gray F. (1992). Accounting and Environmentalism: An Exploration of The Challenge of Gently Accounting For Accountability, Transparency and Sustainability. **Accounting, Organization and Society**, 17 (5) pp. 399-425.

Gray R. (2000). Social and Environmental Responsibility, Sustainability and Accountability: Can the Corporate Sector Deliver? September, CSEAR, Glasgow University, cited in http: //www. gla.ac.uk/ departments/accounting/csear.

GRI (Global Reporting Initiative). (2002). Sustainability reporting guidelines. Amsterdam: GRI.

Gurvitsh, N. and Sidorova, I. (2012). Environmental and Social Accounting Disclosures as a Vital Component of Sustainability Reporting Integrated into Annual Reports of the Baltic Companies for the Years 2007-2011: Based on Companies Listed on NASDAQ OMX Baltic Main List as of June 2012. **GSTF Business Review (GBR**), *2*(1), p.38.

Gurvitsh, N. and Sidorova, I. (2012). Survey of sustainability reporting integrated into annual reports of Estonian companies for the years 2007-2010: based on companies listed on Tallinn Stock Exchange as of October 2011. **Procedia Economics and Finance**, *2*, pp.26-34.

Guthrie, J. and Farneti, F. (2008). GRI sustainability reporting by Australian public sector organizations. **Public Money and management**, *28*(6), pp.361-366.

Hackston D. and Milne, M. (1996). Some determinants of social and environmental disclosure in New Zealand. **Accounting, Auditing and Accountability Journal**, 9 (1) pp. 77-108.

Hahn, R. and Kühnen, M. (2013). Determinants of sustainability reporting: a review of results, trends, theory, and opportunities in an expanding field of research. **Journal of Cleaner Production**, 59 pp. 5-21.

Hedberg, C. J., and Von Malmborg, F. (2003). The global reporting initiative and corporate sustainability reporting in Swedish companies.

Corporate social responsibility and environmental management, 10(3), pp. 153-164.

Hedberg, C. J., von Malmborg, F. (2003). The Global Reporting Initiative and corporate sustainability reporting in Swedish companies. **Corporate Social Responsibility and Environmental Management** 10 pp. 153-164.

Hilson, G. and Basu, A.J. (2003). Devising indicators of sustainable development for the mining and minerals industry: An analysis of critical background issues. **The International Journal of Sustainable Development & World Ecology,** 10(4), pp.319-331.

Hilson, G. and Basu, A.J. (2003). Devising indicators of sustainable development for the mining and minerals industry: An analysis of critical background issues. **The International Journal of Sustainable Development & World Ecology,** 10(4), pp.319-331.

Hilson, G. and Potter, C., (2003). Why is illegal gold mining activity so ubiquitous in rural Ghana? **African Development Review,** 15(2-3), pp.237-270.

Ho, S. S. M. Wong, K. S. (2001). A study of the relationship between corporate governance structures and the extent of voluntary disclosure. **Journal of International Accounting, Auditing and Taxation,** 10 pp. 139–156.

Ho, S.M. and Wong, K.S. (2001). A study of the relationship between corporate governance structures and the extent of voluntary disclosure. Journal of International Accounting, Auditing and Taxation, Vol. 10, pp. 139-56

Hossain, M. and Adams, M. (1995). Voluntary financial disclosure by Australian listed companies. **Australian Accounting Review,** 5(10), pp.45-55.

Hossain, M., Perera, M. H. B. and Rahman, A. R. (1995). Voluntary Disclosure in the Annual Reports of New Zealand Companies. **Journal of International Financial Management & Accounting,** 6, pp. 69–87. doi: 10.1111/j.1467-646X.1995.tb00050.x

IFAC, (2006) IFAC. International Federation of Accountants' handbook of international auditing, assurance, and ethics pronouncements (2006)

IFAD (2007). Private Sector: Development and Partnership Strategy. Rome: IFAD. Retrieved from: http://www.ifad.org/pub/policy/private/e.pdf

IISD (International Institute for Sustainable Development). 1991. Business strategy for sustainable development. Winnipeg, Manitoba, 116 pp.

IISD (International Institute for Sustainable Development). 1992. Trade and sustainable development. Winnipeg, Manitoba.

Institute of Chartered Accountants Ghana, 2000

Isaksson, R., and Steimle, U. (2009). What does GRI-reporting tell us about corporate sustainability? **The TQM Journal**, 21(2), pp. 168-181.

Isaksson, R.,and Garvare, R. (2003). Measuring sustainable development using process models. **Managerial Auditing Journal**, 18(8), pp. 649-656.

Jenkins, H. M. and Yakovleva, N. (2006). "Corporate social responsibility in the mining industry: Exploring trends in social and environmental disclosure". **Journal of Cleaner Production,** 14(3-4), pp. 271-284.

Jensen, M., and W. Meckling, (1979). Rights and production functions: an application to Labour-managed firms. **Journal of Business**, 52, pp. 469–506.

Joseph, G. (2012). Ambiguous but tethered: An accounting basis for sustainability Reporting. **Critical Perspectives on Accounting,** 23 pp. 93– 106.

Kazi, T. G., Afridi, H. I., Kazi, N., Jamali, M. K., Arain, M. B., Jalbani, N., and Kandhro, G. A. (2008). Copper, chromium, manganese, iron, nickel, and zinc levels in biological samples of diabetes mellitus patients. **Biological Trace Element Research**, 122(1), pp. 1-18.

Khalid, S. M, Hennell, A, and Solomon, J. (2013). Site-specific and geographical segmental social, environmental and ethical disclosures by the mining sector. Unpublshed manuscripts.

Kisenyi, V. and Gray, R. H. (1998). Social disclosure in Uganda. **Social and Environmental Accounting**, 18 (2), pp. 16-18.

KPMG (2002). **International Survey of Corporate Sustainability Reporting**, KPMG, Amsterdam.

KPMG (2005). **International Survey of Corporate Responsibility Reporting**. Amsterdam: KPMG.

KPMG (2008). **International Survey of Corporate Sustainability Reporting**. Amsterdam: KPMG.

KPMG (2008a). **KPMG International Survey of Corporate Responsibility Reporting**, KPMG International, Amsterdam.

KPMG (2008b). **Sustainability Reporting: A Guide**, KPMG and Group of 100 (G100), Australia.

Kraut, R.E., Resnick, P., Kiesler, S., Burke, M., Chen, Y., Kittur, N., Konstan, J., Ren, Y. and Riedl, J. (2012). **Building successful online communities: Evidence-based social design**. Mit Press.

Lamberton G. (2005). Sustainable Sufficiency – an Internally Consistent Version of Sustainability. **Sustainable Development**, 13, pp. 53–68.

Lamberton, G. (2005). Sustainability accounting – A brief history and conceptual framework. **Accounting Forum**, 29 (1), pp. 7 – 26.

Langer, M. (2006) **Comparability of sustainability reports. A comparative content analysis of Austrian sustainability reports**. Quoted in: Schaltegger, S., Bennett, M.,

Mantey, J., (2007). College majors could cause women to earn less. **US News and World Report.**

Mashayekhi, B. and Mashayekh, S. (2008). Development of accounting in Iran. **The International Journal of Accounting,** 43 (2008), pp. 66–86.

Matten, D. and Moon, J. (2008). "Implicit" and "explicit" CSR: A conceptual framework for a comparative understanding of corporate social responsibility. **Academy of management Review,** *33*(2), pp.404-424.

Mautz, R.K. (1968). **Financial reporting by diversified companies.** Financial Executives Research Foundation.

Mensah, Y.M., Song, X. and Ho, S.S., (200)3. The long-term payoff from increased corporate disclosures. **Journal of Accounting and Public Policy,** *22*(2), pp.107-150.

Mitchell, R.K., Agle, B.R. and Wood, D.J. (1997). Towards a theory of stakeholder identification and salience: defining the principle of who and what really counts. **Academy of Management Review** 22 (4) pp. 853-886.

Moneva, J. M., Archel, P. and Correa, C. (2006). GRI and the camouflaging of corporate unsustainability. **Accounting Forum** 30, pp. 121-137.

Monteiro, S. M., and Aibar-Guzmán, B. (2010). Determinants of environmental disclosure in the annual reports of large companies operating in Portugal. **Corporate Social Responsibility and Environmental Management,** 17 (4), pp.185-204.

Moore, G. and Robson, A. (2002). The UK supermarket industry: an analysis of corporate social and financial performance. **Business Ethics: A European Review,** *11*(1), pp.25-39.

Nordheim, E. and Barrasso, G. (2007). Sustainable development indicators of the European aluminium industry. **Journal of Cleaner Production** 15 (3) pp. 275-279.

O'Connell, A. A. (2005). **Logistic regression models for ordinal response variables.** Thousand Oaks, CA: Sage Publications.

Radebaugh, L.H. and Gray, S.J., (1993). **Pressures for international accounting harmonisation and disclosure.** International Accounting and Multinational Enterprises, third edition. New York: Wiley.

Rahman Belal, A. (2001). A study of corporate social disclosures in Bangladesh. **Managerial Auditing Journal,** *16*(5), pp.274-289.

Rahman Belal, A. and Owen, D.L. (2007). The views of corporate managers on the current state of, and future prospects for, social reporting in Bangladesh: An engagement-based study. **Accounting, Auditing & Accountability Journal,** *20*(3), pp.472-494.

Ratanajongkol, S., Davey, H. and Low, M. (2006). Corporate social reporting in Thailand: the news is all good and increasing. **Qualitative Research in Accounting and Management** 3 (1) pp. 67-83.

Ratanajongkol, S., Davey, H., and Low, M. (2006). Corporate social reporting in Thailand: The news is all good and increasing. **Qualitative Research in Accounting & Management,** 3(1), pp. 67-83.

Reynolds, M., and Yuthas, K. (2008). Moral discourse and corporate social responsibility reporting. **Journal of Business Ethics,** 78(1-2), pp. 47-64.

Riahi-Belkaoui, A. (2001). The Role for Corporate Reputation for Multinational Firms: **Accounting, Organizational, and Market Considerations**. Westport, CT: Quorum.

Richardson, H. S. (1997). **Democratic intentions. The Modern Schoolman,** 74(4), 285-300.

Robertson, D.C. and N. Nicholson (1996). Expressions of Corporate Social Responsibility Quoted in Robson, C. (2002) **Real world research: A resource for social scientists and practitioner-researchers**. 2nd ed. Oxford: Blackwell.

Roca, L. C. and C. Searcy (2012). An analysis of indicators disclosed in corporate sustainability reports. **Journal of Cleaner Production** 20 pp. 103-118.

Rosa, F. S., Ensslin, S. R., Ensslin, L., and Lunkes, R. J. (2012). Environmental disclosure management: a constructivist case. **Management Decision,** 50 (6), pp. 1117-1136.

Sanders, J., Alexander, S. and Clark, S. (1999). New segment reporting. Is it working?, **Strategic Finance,** 81(6), pp. 35–38.

Santos, J. (2003). E-service quality: a model of virtual service quality dimensions. **Managing Service Quality: An International Journal,** *13*(3), pp.233-246.

Santos, J.A. (2001). Bank capital regulation in contemporary banking theory: A review of the literature. **Financial Markets, Institutions & Instruments,** *10*(2), pp.41-84.

Schaltegger, S., Müller, K. and Hindrichsen, H. (1996). **Corporate environmental accounting**. Chichester: Wiley.

Schiller, H. I. (1981). **Who knows: Information in the age of the Fortune 500**. ABLEX Publishing Corporation.

Searcy, C., McCartney, D. and Karapetrovic, S. (2007). Sustainable development indicators for the transmission system of an electric utility.

Corporate Social Responsibility and Environmental Management 14 (3) pp. 135-151.

Senbet, L.W. and Otchere, I. (2006). Financial sector reforms in Africa: perspectives on issues and policies. In Annual World Bank conference on development economics: Growth and integration, pp. 81-120.

Skouloudis, A. and Evangelinos, K.I. (2009). Sustainability reporting in Greece: are we there yet? **Environmental Quality Management,** 19 (1) pp. 43-59.

Skouloudis, A., Evangelinos, K. and Kourmousis, F. (2010). Assessing non-financial reports according to the Global Reporting Initiative guidelines: evidence from Greece. **Journal of Cleaner Production**, 18 pp. 426–438.

Skouloudis, A., Evangelinos, K.I. and Kourmousis, F. (2009). Development of an evaluation methodology for triple bottom line reports using international standards on reporting. **Environmental Management,** 44 pp. 298-311.

Slater, A. (2008). KPMG International Survey of Corporate Responsibility Reporting 2008. KPMG Global Sustainability Services, The Netherlands.

Smith C. W. Jr. and Watts, R.L. (1982). Incentive and tax effects of U.S. executive compensation plans. **Australian Journal of Management**, 7, pp. 139–157.

Sobhani, F.A., Amran, A. and Zainuddin, Y. (2009). Revisiting the practices of corporate social and environmental disclosure in Bangladesh. **Corporate Social Responsibility and Environmental Management,** 16 pp. 167-183.

Sobhani, F.A., Amran, A. and Zainuddin, Y. (2012). Sustainability disclosure in annual reports and websites: a study of the banking industry in Bangladesh. **Journal of Cleaner Production,** *23*(1), pp.75-85.

SRC Consult (2010). Development of CSR Guidelines for Mining Companies. Minerals Commission of Ghana.

Stanny, E., and Ely, K. (2008). Corporate environmental disclosures about the effects of climate change. **Corporate Social Responsibility and Environmental Management,** 15(6), pp. 338-348.

Steurer, R. (2005). Mapping stakeholder theory anew: from the 'stakeholder theory of the firm' to three perspectives on business–society relations. **Business Strategy and the Environment,** 15(1), pp. 55-69.

Steurer, R., Langer, M. E., Konrad, A., and Martinuzzi, A., (2005). Corporations, stakeholders and sustainable development I: a theoretical exploration of businesse society relations. **Journal of Business Ethics,** 61, pp. 263-281.

Stevens W. P. (1982). The Market Reaction to Corporate Environmental Performance, a paper presented at the American Accounting Association Annual Conference, San Diego. Cited in Mathews M. I. (1993).

Stewart J. D. (1984). The role of information in public accountability in Issues in Public Sector Accounting, Hopwood A. And C. Tomkins (eds) Oxford: Philip Allen. Stock Exchange Yearbook 1994-1995, The Macmillan.

Stiller, Y. and Daub, C. H. (2007). Paving the way for sustainable communication: evidence from a Swiss study. **Business Strategy and the Environment**, 16 pp. 474-486.

Stratos, (2008). Canadian Corporate Sustainability Reporting: Best Practices 2008. Stratos, Toronto.

Suchman, M.C. (1995) Managing legitimacy: strategic and institutional approaches. **Academy of Management Review**, 20 (3) pp. 571-610.

Tonkin, D. and Skerrat, L. (1989). **Financial Reporting: A Survey of Published Accountants.** ICAEW, London.

Toppinen, A., Li, N., Tuppura, A. and Xiong, Y. (2012). Corporate responsibility and strategic groups in the forest-based industry: Exploratory analysis based on the Global Reporting Initiative (GRI) framework. **Corporate Social Responsibility and Environmental Management,** *19*(4), pp.191-205.

Unerman, J. (2000). Methodological issues - Reflections on quantification in corporate social reporting content analysis. **Accounting, Auditing & Accountability Journal**, 13 (5), pp.667 – 681.

Unerman, J. and Bennett, M. (2004). Increased stakeholder dialogue and the internet: towards greater corporate accountability or reinforcing capitalist hegemony? **Accounting, Organisations and Society**, 29 (7) pp. 685-707.

Utaminingtyas, T. H. and Ahalik. (2010). The Relationship Between Corporate Social Responsibility and Earnings Response Coefficient: Evidence from Indonesian Stock Exchange. Oxford Business & Economic Program.

van Berkel, R. (2007). Cleaner production and eco-efficiency initiatives in Western Australia 1996–2004. **Journal of Cleaner Production**, 15(8), pp. 741-755.

Van Berkel, R. and Bossilkov, A. (2004). Sustainable development reporting in the Australian minerals processing industry. **Green Processing**, *2004*, pp.185-195.

Van Marrewijk, M. (2003). Concepts and definitions of CSR and corporate sustainability: Between agency and communion. **Journal of business ethics**, 44(2-3), pp.95-105.

van Staden, and Hooks, J. (2007). A comprehensive comparison of corporate environmental reporting and responsiveness. **British Accounting Review**, 39 (3) (2007), pp. 197–210.

VanMarrewijk, M. (2003). Concepts and definitions of CSR and corporate sustainability: between agency and communion. **Journal of Business Ethics**, 44 (2) pp. 95-105.

Vormedal, I., and Ruud, A. (2009). Sustainability reporting in Norway–an assessment of performance in the context of legal demands and socio-political drivers. **Business Strategy and the Environment**, 18(4), pp. 207-222.

Waddock, S.A., and S.B. Graves. (1994) Industry Performance and Investment in R&D and Capital Goods. **The Journal of High Technology Management Research**, 5 (1), pp. 1-17.

Wagner, M., Phu, N. G., Azomahou, T., and Wehrmeyer, W. (2002). The relationship between the environmental and economic performance of firms: an empirical analysis of the European paper industry. **Corporate Social Responsibility and Environmental Management,** 9 (3), pp. 133-146.

Wang, K., Sewon, O. and Claiborne, M. C. (2008). Determinants and consequences of voluntary disclosure in an emerging market: Evidence from China. **Journal of International Accounting, Auditing and Taxation**, 17 (1) pp. 14–30.

Watson, A., Shrives, P., and Marston, C. (2002). Voluntary disclosure of accounting ratios in the UK. **British Accounting Review**. 34, (4), pp. 289-313

Watts, R. L. and Zimmerman, J.L. (1990). Positive accounting theory: A ten year perspective. **The Accounting Review**, 65 (1), pp. 131–156.

Weber, R. (1990). **Quantitative applications in the social sciences: Basic content analysis**. 2nd ed. Beverly Hills, Calif: Sage Publications.

Weber, R.P. (1990). **Basic Content Analysis, Sage University Paper Series on Quantitative Applications in the Social Sciences**, Series No. 49 2nd ed. Newbury Park: Sage Publications.

White, H. (1980). A heteroskedasticity-consistent covariance matrix and a direct test for heteroskedasticity. **Econometrica**, 48, pp. 817-838.

Wilks, I. (1989). Asante in the nineteenth century: Structure and evolution of a political order. (2nd Ed.) Cambridge University Press, London (1989

Williams, S.M. and Tower, G. (1998). Differential reporting in Singapore and Australia: A small business managers' perspective. **The International Journal of Accounting**, *33*(2), pp.263-268.

Willis, A. (2003). The role of the global reporting initiative's sustainability reporting guidelines in the social screening of investments. **Journal of Business Ethics**, 43 (3), pp. 233-7.

Wilmshurst, T. D. and Frost, G. R. (2000). Corporate environmental reporting: A test of legitimacy theory. **Accounting, Auditing & Accountability Journal**, 13 (1), pp.10 – 26.

Wiseman, J. (1982). An evaluation of environmental disclosures made in corporate annual reports. **Accounting, Organizations and Society**, 7 (1), pp. 53-63.

Woodward, D., Edwards, P. and Birkin, F. (2001). Some evidence on executives' views of corporate social responsibility. **British Accounting Review**. 33 (3), pp. 357-397.

World Business Council for Sustainable Development (WBCSD), (2002). Sustainable Development Reporting: Striking the Balance. World Business Council for Sustainable Development, Geneva.

World Business Council for Sustainable Development. WBCSD. (2002). Sustainable development reporting: striking the balance. Hertfordshire: Earth print.

World Commission on Environment and Development (WCED), (1987). **Our Common Future**. Oxford University Press, Oxford, UK.

Yin, R.K. (2009). **In: Case Study Research: Design and Methods**. 4th ed. SAGE Publications, Thousand Oaks, California, USA.

Zarzeski MT. (1996). Spontaneous harmonization effects of culture and market forces on accounting disclosure practices. **Accounting Horizons** 10(1), pp. 18–37.

Zeghal S. D. and Ahmed, A. (1990). Comparison of Social Responsibility Information Disclosure Media Used by Canadian Firms. **Accounting, Auditing & Accountability Journal**, 3 (1) pp. 38 49.

Zeng, S. X., Hu, X. D., Yin, H. T., and Tam, C. M. (2012). Factors that drive Chinese listed companies in voluntary disclosure of environmental information. **Journal of Business Ethics**, 109 (3) pp. 309-321.

Ziegler, A. and Schröder, M. (2010). What determines the inclusion in a sustainability stock index? A panel data analysis for European firms. **Ecological Economics** 69, pp. 848-856.